# Introductory Physics of Nuclear Medicine

# Introductory Physics of Nuclear Medicine

Ramesh Chandra, Ph.D.

*Professor*
*Department of Radiology*
*New York University*
*Medical School*

Fourth Edition

*Lea & Febiger* Philadelphia • 1992

Lea & Febiger
200 Chester Field Parkway
Malvern, Pennsylvania 19355-9725
U.S.A.
(215) 251-2230

Executive Editor—Darlene Cooke
Project/Manuscript Editor—Jessica Howie Martin
Production Manager—Samuel A. Rondinelli

**Library of Congress Cataloging-in-Publication Data**

Chandra, Ramesh, 1938—
Introductory physics of nuclear medicine / Ramesh Chandra. — 4th ed.
p. cm.
Includes bibliographical references and index.
ISBN 0-8121-1442-6
1. Medical physics. 2. Nuclear medicine. 3. Radioisotopes. I. Title.
[DNLM: 1. Nuclear Medicine. 2. Physics. WN 415 C4561]
R895.C47 1992
616.07′575—dc20
DNLM/DLC
for Library of Congress 91-46717
CIP

Fourth Edition, 1992

Reprints of chapters may be purchased from Lea & Febiger in quantities of 100 or more. Contact Sally Grande in the Sales Department.



PRINTED IN THE UNITED STATES OF AMERICA

Print number: 5 4 3 2 1

*To my wife, Mithilesh,*
*my son, Anurag,*
*and my daughter, Ritu*

# *Preface*

The reception of the present book by resident physicians in radiology and nuclear medicine as well as by students of nuclear medicine technology has reinforced my belief in the need for an introductory book on physics and related basic sciences of nuclear medicine. Therefore, the purpose, the audience, and the level of subject matter remain the same as in previous editions. I quote, "This book is primarily addressed to resident physicians in nuclear medicine as well as residents in radiology, pathology and internal medicine, who wish to acquire some knowledge of nuclear medicine. Nuclear medicine technicians wishing to advance in their field should also find this book useful.

"I have tried to write in a simple and concise manner, including not only essential details, but also many examples and problems taken from the routine practice of nuclear medicine. Basic principles and underlying concepts are explained. Mathematical equations or in some cases their derivations have been included only when it was felt their inclusion would help the reader and when it was essential for the proper development of the subject matter. However, the reader is warned that this is not an introductory book of physics and, therefore, some familiarity on his or her part with the elementary concepts of physics such as units, energy, force, electricity, and light is assumed by the author."

Keeping the level and spirit of the earlier editions intact, four major alterations have been made in this edition.

1. A problem set has been added at the end of each chapter. The problems are designed to emphasize the salient features and concepts presented in each chapter and to enhance the learning experience of the student by making him or her think and analyze the problem before arriving at the solution.
2. SI units have been given equal footing to the old system (CGS) of units. Unfortunately, we are still in a transition period in which both units are being used. Therefore, old units could not be entirely removed.
3. New radiopharmaceuticals, particularly for heart and brain imaging, have been included.
4. Radiation doses from common radiopharmaceuticals, radiation effects, and rules and regulations governing the usage of radionuclides have been updated.

In addition to these changes, there are minor revisions and additions throughout the book.

I thank all those people who wrote to me either commending the book or pointing out the errors and omissions in previous editions. Both groups helped me in the preparation of this edition; the first group by boosting my morale and the second group by their suggestions. Individually, I thank Dr. D. Pizzarello who always made himself available for discussions and served as a sounding board for my ideas. Ms. Martha Rosaly labored hard in typing and retyping the manuscript; many thanks to her. Last, but not least, I thank the publisher and his staff for their help and cooperation in bringing my labor to fruition.

*New York, New York* Ramesh Chandra

# Contents

# 1

# Basic Review

From a physicist's point of view, nature consists mainly of matter and the forces governing the behavior of matter. This chapter reviews briefly some of the aspects of the atomic structure of matter which are essential for the understanding of subsequent subject matter.

## MATTER, ELEMENTS AND ATOMS

All matter is composed of a limited number of elements (105 identified so far) which in turn are made of atoms. An atom is the smallest part of an element which retains all the chemical properties of that element in bulk. In general, atoms are electrically neutral; that is, they do not show any electric charge. However, atoms are not indivisible as once was thought, but are composed of three elementary particles—electrons, protons and neutrons.

An electron is a tiny particle which possesses a negative charge of $1.6 \times 10^{-19}$ Coulomb (unit of charge) and a mass of $0.9109 \times 10^{-27}$ gm. A proton is a particle with a positive charge equal in amount to that of an electron. A neutron does not have any electric charge and weighs slightly more than a proton. Protons and neutrons have masses of $1.6726 \times 10^{-24}$ and $1.6747 \times 10^{-24}$ gm, respectively; hence they are about 2,000 times heavier than an electron.

## SIMPLIFIED STRUCTURE OF AN ATOM

An atom is generally neutral because it contains the same number of electrons and protons. The number of protons in an atom is also known as the atomic number Z. It specifies the position of that element in the periodic table and, therefore, specifies its chemical identity. The electrons, protons and neutrons in an atom are arranged in a planetary structure in which the protons and neutrons (the sun) are located at the center and the electrons (planets) are revolving over the surface of spherical shells (or orbits) of different radii. The center in which the protons and neutrons are located is

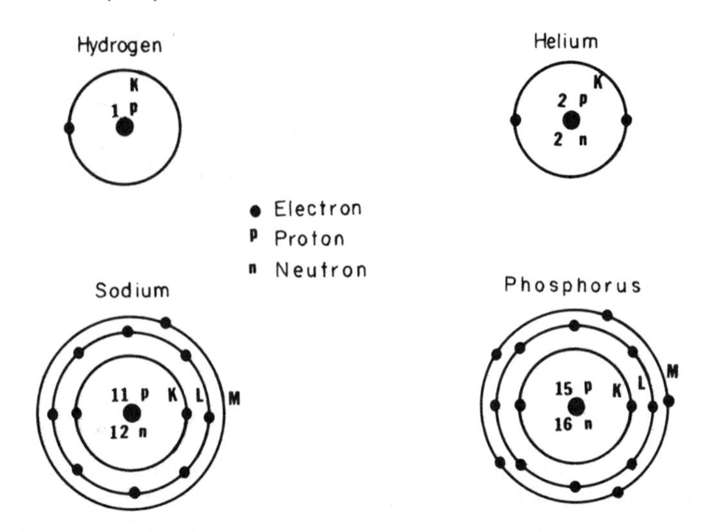


**Fig. 1–1.** Simplified atomic structure of four elements in their ground state.

known as the nucleus and is similar to a packed sphere. The size of atoms of different elements varies greatly but is in the range of 1–2 $\times$ $10^{-8}$ cm. The nucleus is small in comparison to the atom (*i.e.*, about $10^5$ times smaller or $10^{-13}$ cm in size).

The attractive coulomb (electrical) force between the positively charged nucleus (due to the protons) and the negatively charged electrons provides stability to the electrons revolving in the spherical shells. The first shell (having the smallest radius) is known as the K shell, the second shell as L, third shell as M, and so on. There is a limit to the number of electrons which can occupy a given shell. The K shell can be occupied by a maximum of 2 electrons, the L shell by a maximum of 8 electrons, the M shell by a maximum of 18, and the N shell by a maximum of 32 electrons. However, the outermost shell in a given atom cannot be occupied by more than 8 electrons. In a simple atom like hydrogen, there is only one electron which under normal circumstances occupies the K shell. In a complex atom like iodine, there are 53 electrons which are arranged in the K, L, M, N and O orbits in numbers of 2, 8, 18, 18, and 7, respectively. The arrangement of electrons in various shells for hydrogen and three other typical atoms is shown pictorially in Fig. 1–1. This is a simplified description of the atomic structure which, in reality, is more complex as each shell is further divided into subshells, etc. For our purpose, though, it is more than sufficient.

## MOLECULES

Molecules are formed by the combination of 2 or more atoms, *e.g.*, a molecule of water, $H_2O$, is formed by the combination of 2 hydrogen atoms and 1 oxygen atom. The combination of atoms is accomplished through the interaction of electrons (also known as valent electrons) in the outermost orbits of the atom. Valent electrons participate in the formation of the molecules in several ways—for example, in ionic binding, covalent binding, and hydrogen binding. In theory, the majority of chemical reactions and chemical properties of atoms or molecules can be explained on the basis of the interaction of the valent electrons.

## BINDING ENERGY, IONIZATION AND EXCITATION

Each electron in a given shell is bound to the nucleus with a fixed amount of energy. Therefore, if one wishes to remove an electron from a particular shell, make it free and no longer associated with that atom, energy will have to be provided to the electron from outside the atom. The minimum amount of the energy necessary to free an electron from an atom is known as the binding energy of the electron in that atom. The unit in which energy is measured on the atomic scale is known as an electron volt (eV) which is the energy acquired by an electron accelerated through 1 volt of potential difference. The electrons in the K shell are the most tightly bound electrons in an atom and, therefore, require the most energy to be removed from the atom (removal of an electron from an atom is called ionization). Electrons in the outermost shell, on the other hand, are the least tightly bound electrons and, therefore, require the least amount of energy for their removal from the atom. The binding energy of electrons in various shells increases rapidly with the atomic number Z. Table 1–1 lists the K- and L-shell average binding energies of electrons in the atoms of various elements.

Under normal conditions, electrons occupy the lowest possible shells (those closest to the nucleus) consistent with the maximum number of elec-

**Table 1–1. Average Binding Energies of K and L Shell Electrons in Various Elements**

| Element | Atomic Number Z | Average Binding Energy (keV) K Shell | Average Binding Energy (keV) L Shell |
|---|---|---|---|
| H | 1 | 0.014 | — |
| C | 6 | 0.28 | 0.007 |
| O | 8 | 0.53 | 0.024 |
| P | 15 | 2.15 | 0.19 |
| S | 16 | 2.47 | 0.23 |
| Fe | 26 | 7.11 | 0.85 |
| Zn | 30 | 9.66 | 1.19 |
| Br | 35 | 13.47 | 1.78 |
| As | 47 | 25.51 | 3.81 |
| I | 53 | 33.17 | 5.19 |
| In | 69 | 59.40 | 10.12 |
| W | 74 | 69.52 | 12.10 |
| Pb | 82 | 88.00 | 15.86 |

trons by which a given shell can be occupied. However, electrons can be made to move into higher shells (unoccupied shells) temporarily by the absorption of energy. This absorption can take place in various ways—for example, by heating a substance, subjecting matter to high electric fields, passage of a charged particle through matter, or even by a high mechanical impact. When an electron absorbs sufficient energy for its removal from the atom, the process is called ionization and the remaining atom, an ion. When the electron absorbs amounts of energy which are just sufficient to move it into a higher unoccupied shell, the process is known as excitation and the atom as an excited atom. Excited atoms are, in general, unstable and acquire their normal configuration by emitting electromagnetic radiation (light, ultraviolet [UV] light or x-rays), generally within $10^{-9}$ seconds.

For example, let us consider a sodium atom, which has an atomic number of 11 and, therefore, 11 electrons and 11 protons. The electrons are arranged in K, L and M shells in numbers of 2, 8 and 1, respectively. The energies of these electrons in the K, L and M shells are approximately −1072, −63 and −1 eV, respectively. To remove an electron from the K shell of a sodium atom, it is necessary to provide an amount of energy equal to 1072 eV, whereas from the M shell only 1 eV of energy is necessary. An electron from the L shell can move to the M shell by absorbing 62 eV of energy, thereby producing an excited atom of sodium. When this excited atom decays (*i.e.*, when the electron jumps back into the L shell), an electromagnetic radiation of 62 eV will be emitted.

It should be pointed out here, that in the case of electrons, the zero of energy scale, by convention, is chosen to be the state when the electron is just free, and not bound to the atom. As a result, when an electron is bound in the atom, its energy is represented as negative and when an electron is free and moving (*i.e.*, possesses kinetic energy), its energy is represented as positive. This convention is different than the one used in the case of nucleus discussed in the next chapter (p. 21).

## FORCES OR FIELDS

Force is a general term related to the interaction of various constituents of matter. At present four kinds of forces (or fields) are known: gravitational, weak, electromagnetic and strong. Gravitational forces are produced as a result of the mass of matter and play a significant role in holding our solar system intact, but they are negligible between atoms and molecules. Weak forces play a significant part in nuclear transformation which will be described in Chapter 2. Electromagnetic forces play a dominant role in our daily life, since they hold the atom together and are responsible for interactions between atoms, molecules, biomolecules, etc. Strong forces are the forces which hold a nucleus together and act between proton-proton, proton-neutron and neutron-neutron. The relative strengths of these forces are listed below.

| *Type of Force* | *Strength* |
|---|---|
| Strong | 1 |
| Electromagnetic | $10^{-2}$ |
| Weak | $10^{-13}$ |
| Gravitational | $10^{-39}$ |

## Electromagnetic Forces

Electromagnetic forces or fields are produced by charged particles. During interaction between charged particles, quite often energy is emitted as electromagnetic radiation. Electromagnetic radiation can propagate either as waves or as particles. When electromagnetic radiation behaves like particles, these particles are called photons. A photon does not have any rest mass or charge. It is a packet of energy which interacts with matter in a specified manner or according to the laws of electromagnetic forces. The dual nature of radiation, which is now an established fact, is true of matter (*e.g.*, electrons) as well. Electromagnetic radiation is characterized by its energy or wavelength only. Electromagnetic radiation of varying energies is known by different names (shown in Fig. 1–2). The energy of electromagnetic radiation is related to the wavelength by a simple relationship:

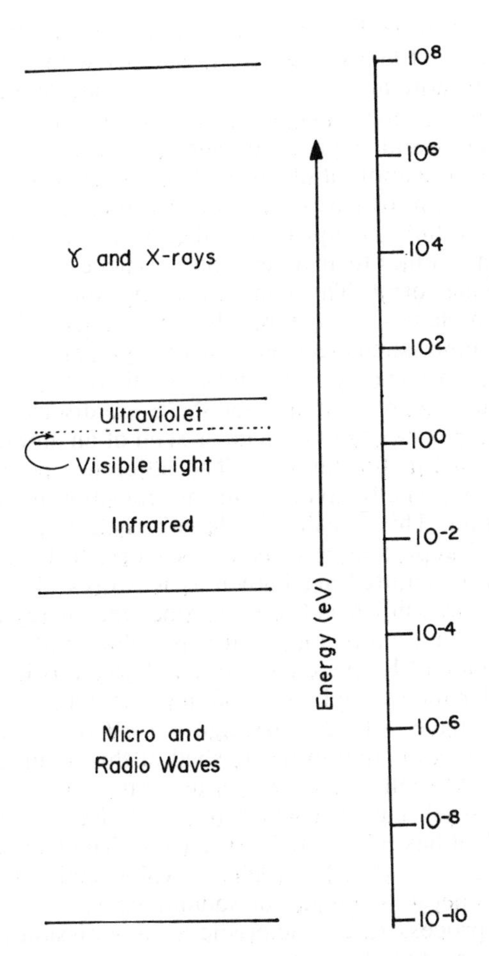


**Fig. 1–2.** Spectrum of electromagnetic waves. Electromagnetic waves of different energies are known by various names. For example, x- or γ-waves are electromagnetic waves with energies higher than 100 eV.

$$E = \frac{hc}{\lambda}$$

where h is Planck's constant, c is the velocity of light or electromagnetic radiation and λ is the wavelength. The above relationship is further reduced if one measures energy in keV and the wavelength in angstroms (Å = $10^{-8}$ cm):

$$E\ (\text{keV}) = \frac{12.4}{\lambda}$$

## CHARACTERISTIC X-RAYS AND AUGER ELECTRONS

As can be seen from Fig. 1–2, x-rays are part of the electromagnetic radiation spectrum. Electromagnetic radiations with energies of approximately 100 eV or more are called x-rays. X-rays are primarily distinguished from other forms of electromagnetic radiation by their ability to produce ionization in matter and to penetrate substances. Characteristic x-rays are produced by the transition of electrons from outer to inner orbits of atoms (the K or L shell in most cases). The inner orbits of an atom under normal circumstances are fully occupied and, therefore, to cause the transition of an electron from an outer to an inner orbit, it is necessary to create a vacancy or hole in the inner orbit. This can be accomplished in various ways. The well-known example is an x-ray tube where high-energy electrons sometimes collide with the inner-shell electrons, knocking them out of the target atoms and thus creating a vacancy in the inner shells of the target atoms. Other examples of vacancy creation in inner shells are discussed in Chapter 2.

Once a vacancy is created in the inner shell of an atom, an electron from the outer shell falls to fill this hole. The difference between the potential energies of the two shells involved in the transition is emitted as electromagnetic radiation. This is called a characteristic x-ray of the atom if the energy emitted is approximately 100 eV or more. If the vacancy was in the K shell, the x-rays emitted are known as K x-rays; if the vacancy was in the L shell, they are known as L x-rays. Since the energy of a characteristic x-ray emitted by an atom is unique, it is possible to identify an atom (and, therefore, an element) by the energy of its characteristic x-rays.

Let us consider the example of a sodium atom where electrons in the K, L, and M shells have −1072, −63, and −1 eV of energies, respectively. Now if a vacancy is created in the K shell of this atom, it will be filled by one of the L- or M-shell electrons. Let us assume it is filled by the L-shell electron. The L-shell electron which originally had an energy of −63 eV now in the K shell has only −1072 eV energy. The balance of this energy, $e_L - e_K = [-63 - (-1072)] = 1009$ eV, will be emitted as a characteristic K x-ray and its energy is unique for sodium atom.

An alternate process to characteristic x-ray emission is the emission of Auger electrons (named in honor of their discoverer, P. Auger). In this process, the vacancy in the K shell is filled by an electron from the L or M

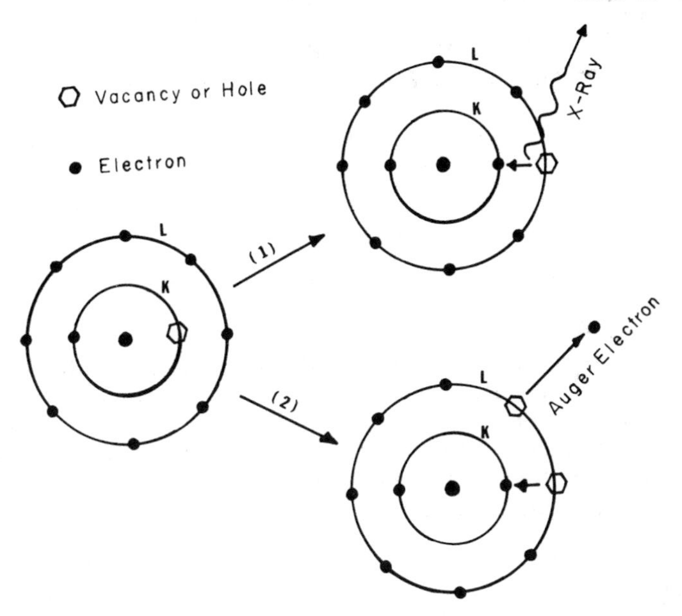


**Fig. 1–3.** X-ray emission and Auger-electron emission. When there is a vacancy (or hole) in the K shell of an atom, one of two processes may occur: either (1) an electron from the L shell (or higher shells) fills the vacancy in the K shell, and the balance of energy between the K and L shells is emitted as a characteristic x-ray; or (2) one of the electrons from the L shell fills the vacancy in the K shell, and the balance of energy is taken by another electron in the L shell (or higher shell) which is then emitted as an Auger electron.

shells; however, the balance of energy which would have been emitted as an x-ray is taken by another electron from the L or M shell. Thus in Auger-electron emission, two electrons are removed from L and/or M shells: one fills the vacancy in K shell, the other is emitted from the atom with the balance of energy. The atom is now doubly ionized. A similar process may occur if the vacancy is in the L or M shell. The Auger process occurs most frequently in elements with low atomic numbers ($Z < 24$, *e.g.*, C, N, O, Al, Ca), whereas x-ray emission is characteristic of elements with high atomic numbers ($Z > 45$, *e.g.*, I, Cs, W, Pb). These two processes, characteristic x-ray emission and Auger-electron emission, are graphically depicted in Fig. 1–3.

## INTERCHANGEABILITY OF MASS AND ENERGY

In 1905 Einstein derived an expression on theoretical grounds which showed the interconvertibility of mass and energy. With subsequent advances in experimental atomic and nuclear physics, his assertion proved to be correct. The expression that relates mass and energy is simply

$$E = mc^2$$

**Table 1–2. Mass and Energy Equivalence for Electrons, Protons and Neutrons**

| *Particle* | *Mass (amu)* | *Mass Energy (MeV)* |
|---|---|---|
| $e^-$ | 0.000549 | 0.511 |
| p | 1.00728 | 938.28 |
| n | 1.00867 | 939.57 |

where E is the energy, m is the mass and c is the velocity of light. Mass on an atomic or nuclear scale is measured in atomic mass units (amu), defined as 1/12 of the mass of a carbon atom. From Einstein's equation for mass and energy conversion, 1 amu is equivalent to 931 MeV of energy. The masses in amu and their equivalent energy in MeV for an electron, proton, and neutron are given in Table 1–2.

## PROBLEMS

1. Find the energies of radiations emitted from the atoms of (a) iodine and (b) lead when there is a vacancy in the K shell of these atoms. (Use Table 1–1.)
2. Determine the minimum energy needed to create a vacancy in (a) the K shells of P, Fe, and In atoms; and (b) the L shells of Zn, I, and Pb atoms.
3. What are the wavelengths of the 10, 100, and 1000 keV x-rays?
4. If the rest mass of an electron is converted into electromagnetic energy, what is the energy of this radiation? What is its name?

# 2

# Nuclides and Radioactive Processes

In the previous chapter I discussed the structure of the atom and pointed out the importance of electrons and their interaction. In this chapter I shall discuss the properties of the nucleus (the central core of an atom) and the radioactive processes which are strictly nuclear phenomena.

## NUCLIDES AND THEIR CLASSIFICATION

As in the case of an atom different types of atoms are called elements, in the case of a nucleus different types of nuclei are termed nuclides. An element is characterized by its atomic number (Z) alone, whereas a nuclide is characterized by its mass number (A) as well as by its atomic number (Z). The mass number (A) of a nuclide is the sum of the number of its protons (Z) and neutrons (N), *i.e.*, $A = Z + N$.

As an example, $^{131}_{53}I$, a nuclide of iodine, contains 53 protons and 78 neutrons which add up to 131, the mass number of this nuclide. A general notation for a nuclide is $^{A}_{Z}X$ where A and Z are, respectively, the mass number and the atomic number of the nuclide, and X is the element to which this nuclide belongs.

Nuclides are classified according to their mass number, neutron number and atomic (proton) number. Nuclides having the same *a*tomic mass are called isob*a*rs, *e.g.*, $^{131}_{53}$ and $^{131}_{54}Xe$ (both these nuclides have the same mass number, 131). Nuclides having the same number of *p*rotons are called Isoto*p*es, *e.g.*, $^{12}_{6}C$ and $^{13}_{6}C$. Since they contain the same number of protons and, therefore, have the same atomic number, isotopes always belong to the same element. Nuclides having the same number of *n*eutrons are called isoto*n*es, *e.g.*, $^{13}_{7}N$ and $^{14}_{8}O$ (both of these nuclides contain 6 neutrons). As an aid to memory, a pair of corresponding letters in each definition have been italicized.

| | |
|---|---|
| same number of *p*rotons | isoto*p*es |
| same *a*tomic mass | isob*a*rs |
| same number of *n*eutrons | isoto*n*es |

## NUCLEAR STRUCTURE AND EXCITED STATES OF A NUCLIDE

How are neutrons and protons (a common name for the two combined is nucleon) arranged inside the nucleus? We have only a partial answer to this question so far. Our understanding of the structure of a nucleus as compared to the structure of an atom (*i.e.*, arrangement of electrons in various shells and their interactions) is quite limited. It is postulated—with a wealth of supportive experimental data—that nucleons are arranged in spherical shells inside the nucleus in a manner similar to that of electrons in an atom. Not much is known, however, of the manner in which nucleons are stacked in these shells or about the transitions of nucleons between different shells. What is known and is important for the present discussion is that nucleons, like electrons in an atom, can also be excited to higher unoccupied shells by absorption of energy from outside the nucleus. The lowest possible arrangement of nucleons in the nucleus is known as the ground state of a nuclide. The higher shells are commonly referred to as energy levels or excited states. Again similar to electrons in an atom, nucleons in a nucleus are also bound with different binding energies. The binding energy of a nucleon (amount of energy needed to pull it out of the nucleus) varies from nuclide to nuclide. However, the average binding energy of nucleons for most nuclides is in the range of 5–8 MeV. This is about 1000 times higher (MeV rather than keV) than the average binding energies of electrons in atoms. Therefore, it is hard to remove a proton or neutron from a nucleus. It requires transfer of a large amount of energy from outside which is, in general, only possible in nuclear reactors, accelerators, or cyclotrons.

The excited states of a nuclide, in general, are of a short duration ($<10^{-11}$ sec) and decay to the ground state or lower energy states by emission of high-energy radiation in a manner similar to that of the excited states of an atom which decay to the ground state of an atom by emitting light or x-rays.

*E*xcited states have the same mass number, same atomic number and the same number of neutrons as the ground state and, therefore, are called isomers; *i.e.*, an isom*e*r is an *e*xcited state of a nuclide. Note that the letter *e* has been italicized twice in the above definition, again as an aid to memory. The isomer of a nuclide is generally distinguished from its ground state by placing an asterisk after the symbol of the nuclide, *e.g.*, ${}^{12}_{6}C^*$ is an excited state of ${}^{12}_{6}C$. In a few cases, however, the lifetime of the excited state of a nuclide can be very long (seconds, minutes or even years). When this occurs, the excited state is called a metastable state. Well-known examples of nuclides having metastable states are ${}^{99m}Tc$ and ${}^{113m}In$. The letter "m" after the mass number in these two nuclides stands for metastable state and distinguishes them from their respective ground states of ${}^{99}Tc$ and ${}^{113}In$.

## RADIONUCLIDES AND STABILITY OF NUCLIDES

Even in their ground state many nuclides are unstable. These unstable nuclides are called radionuclides. Radionuclides try to become stable by emitting electromagnetic radiation or charged particles. Electromagnetic radiation or charged particle emission is called radioactive decay. What makes

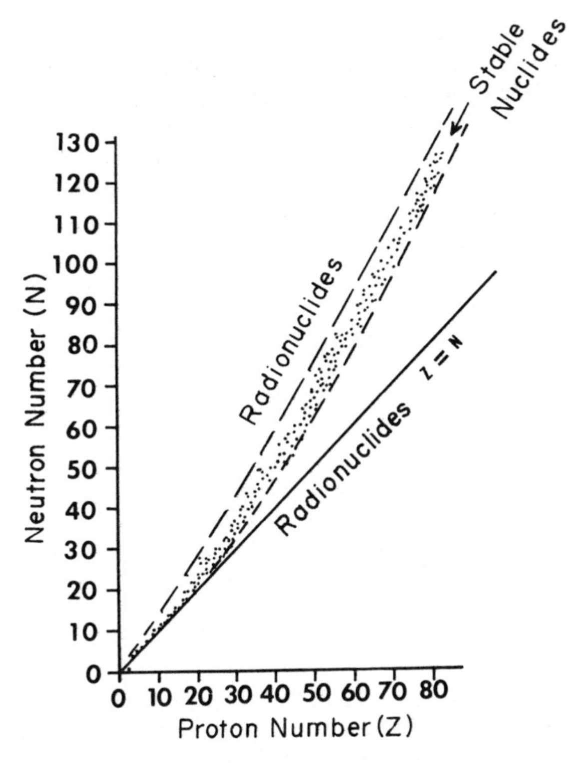


**Fig. 2–1.** Plot of the number of neutrons as a function of number of protons (atomic number) for stable nuclides. In the low atomic-number elements, the number of protons is roughly equal to the number of neutrons, but in the higher atomic-number elements, more neutrons are needed than protons for stability. Radionuclides lie on either side of the stability curve.

a nuclide stable or radioactive? Two kinds of forces, strong and electromagnetic, determine the stability of a nuclide. The strong forces act between a pair of nucleons, *e.g.*, proton-proton, proton-neutron or neutron-neutron. They are attractive and act only when the distance between the two nucleons is very small. Electromagnetic forces act between protons only (because there is no charge on neutrons) and are repulsive (similar charges repel each other). The balance between these two forces—one attractive, the other repulsive—determines the stability of a nuclide. Whenever the balance between these two forces is disturbed, the nuclide becomes unstable and, therefore, radioactive. There are approximately 259 stable nuclides found in nature. If one plots the number of neutrons (N) contained as a function of the number of protons (Z) in a nuclide for all stable nuclides, a curve as shown in Fig. 2–1 results. This curve initially starts as a straight line and then slowly bends toward the neutron number (N) for higher atomic numbers. From this curve it can be seen that for a lighter nuclide ($A < 50$) the number of protons in stable nuclides is equal to the number of neutrons; for example, $^{16}_{8}O$, a stable oxygen nuclide, contains 8 neutrons and 8 protons. For heavier nuclides (A > 100), however, the number of neutrons needed for the stability of a nuclide is much more than the number of protons; for example, $^{127}_{53}I$, a stable iodine nuclide, contains 53 protons and 74 neutrons. The region on either side of the curve in Fig. 2–1 is called the domain of radionuclides. If a radionuclide lies in the upper region, it contains an excess number of neutrons which is causing the unstability of the radionuclide. On the other hand, if the radionuclide lies in the lower region, it is the excess of protons which is making the nuclide unstable.

## RADIOACTIVE SERIES OR CHAIN

A radionuclide tries to attain stability through radioactive decay. The stability may be achieved either by direct (single-step) decay to a stable nuclide, or by decaying to several radionuclides in multiple steps and, finally, to a stable nuclide. For example, $^{131}_{53}I$ directly decays to a stable nuclide $^{131}_{54}Xe$. On the other hand, $^{226}_{88}Ra$ first decays to $^{222}_{86}Rn$, which decays to $^{218}_{84}P$ . . . , and finally to $^{210}_{82}Pb$, a stable nuclide. The complete series for this is given below (the arrows denote the sequence of decay).

$$^{226}_{88}Ra \rightarrow {}^{222}_{86}Rn \rightarrow {}^{218}_{84}Po \rightarrow {}^{214}_{82}Pb \rightarrow {}^{214}_{83}Bi \rightarrow {}^{214}_{84}Po \rightarrow {}^{210}_{82}Pb \text{ (Stable Nuclide)}$$

This latter process is often referred to as a radioactive series or chain. Well-known examples of a radioactive series in nuclear medicine are the decay of $^{99}Mo$ and $^{113}Sn$ which go through the following sequences:

$$^{99}Mo \rightarrow {}^{99m}Tc \rightarrow {}^{99}Tc \rightarrow {}^{99}Ru \text{ (Stable)}$$

$$^{113}Sn \rightarrow {}^{113m}In \rightarrow {}^{113}In \text{ (Stable)}$$

## RADIOACTIVE PROCESSES AND CONSERVATION LAWS

The three processes through which a radionuclide tries to attain stability are called alpha ($\alpha$), beta ($\beta$), and gamma ($\gamma$) decay. These names were given because at the time of the discovery of these processes their exact nature was not known. Three important conservation laws are pertinent here since they always hold true in radioactive processes or nuclear transformations. These are the law of conservation of energy, the law of conservation of mass number, and the law of conservation of electric charge.

The law of conservation of energy states that the total energy (mass energy + kinetic energy + energy in any other form, *e.g.*, a photon) remains unchanged during a radioactive process or nuclear transformation. The law of conservation of mass number states that the sum of mass numbers remains unchanged in radioactive or nuclear processes. The mass number of a neutron or proton is assumed to be one and that of an electron zero. Similarly, the law of conservation of electric charge states that the total charge during a radioactive process or nuclear transformation remains unchanged.

## Alpha Decay

In alpha decay, a radionuclide emits a heavy, charged particle called an $\alpha$ particle. An $\alpha$ particle is 4 times heavier than a proton or neutron and carries an electric charge which is twice that of a proton. In fact, an $\alpha$ particle is a stable nuclide with atomic mass number A = 4 and atomic number Z = 2. This happens to be the nucleus of a helium atom.

From the conservation laws of the mass number and electric charge it follows that during alpha decay the mass number and the atomic number of the resulting nuclide (also known as the daughter nuclide) will be reduced by 4 and 2, respectively. Alpha decay can be expressed by the following equation:

$$^{A}_{Z}X = ^{(A-4)}_{(Z-2)}Y + ^{4}_{2}He \ (\alpha \text{ particle})$$

Notice that the mass number and the electric charge (in this case the sum of atomic numbers) on both sides of the arrow are the same. An example of alpha decay is the decay of radium 226 to radon 222:

$$^{226}_{88}Ra = ^{222}_{86}Rn + \alpha$$

*Example:*

$^{222}_{86}Rn$ is a radionuclide which decays through alpha decay. Determine the nature of its daughter nuclide. Since $^{222}_{86}Rn$ decays through the alpha decay process, the mass number of the daughter nuclide will be 4 less than that of the parent, *i.e.*, 222 − 4 = 218. Similarly, the atomic number will be reduced by 2, *i.e.*, 86 − 2 = 84. Therefore, the daughter nuclide has a mass number 218 and its atomic number is 84. According to the periodic table, the atomic number 84 belongs to the element Po. Therefore, the daughter nuclide is $^{218}_{84}Po$.

Radon-222 is a radioactive gas which escapes into the environment from the rocks or soil containing radium-226. A small amount of radium is present everywhere. Such amounts are generally no more than a nuisance that increases the background radiation slightly. However, a combination of relatively large concentrations of radium in the soil and rocks and poor circulation in a house can sometimes produce hazardous concentrations of radon in the house, which have to be dealt with effectively for the safety of the occupants.

Two salient facts concerning alpha decay to remember are: (1) it occurs mostly with radionuclides whose atomic mass number A is greater than 150; and (2) the kinetic energy of the emitted $\alpha$ particle is fixed and discrete for a given decay. In the above example of $^{226}_{88}Ra$ alpha decay, the kinetic energy of the $\alpha$ particle emitted is 4.780 MeV.

## Beta Decay

During this transformation, a neutron or a proton inside the nucleus of a radionuclide is converted to a proton or a neutron, respectively. When a proton is converted into a neutron, the positive charge inside the nucleus decreases by one, reducing the repulsive forces between protons. On the other hand, when a neutron is converted into a proton, the positive charge inside the nucleus increases by one, and therefore the repulsive forces between protons are increased. The result of the decrease or increase in the repulsive force within the nucleus is to balance the two forces (electromagnetic and strong) in an attempt to stabilize the nuclide (see Radionuclides and Stability of Nuclides, p. 11).

The conversion of a neutron into a proton or of a proton into a neutron is controlled by weak (as opposed to strong, electromagnetic and gravitational) forces (see Chapter 1). The exact nature of weak forces is not known and is not pertinent to our discussion of beta decay. Beta decay occurs through one of the following processes: (1) $\beta^-$ emission or electron emission, (2) $\beta^+$ emission or positron emission, or (3) electron capture.

***$\beta^-$ Emission:*** In this process a neutron inside the radionuclide is converted into a proton and the excess energy is released as a pair of particles, an electron and an antineutrino ($\bar{\nu}$). As such, there are no electrons or antineutrinos inside the nucleus. These are created from excess energy at the instant of radioactive decay. An antineutrino is a particle which has no rest mass and no electric charge ($Z = 0$). It rarely interacts with matter and, therefore, is of no biological significance. Its existence was postulated so that the law of energy conservation would not be violated. However, its existence has now been confirmed experimentally. $\beta^-$ decay can be expressed as:

$$^{A}_{Z}X = {}^{A}_{Z+1}Y + e^- + \bar{\nu}$$

Note that both the mass number and the electric charge are conserved. Some well-known examples of $\beta^-$ decay are $^{3}_{1}H$, $^{14}_{6}C$ and $^{32}_{15}P$, which can be expressed as follows:

$$^{3}_{1}H = {}^{3}_{2}He + e^- + \bar{\nu}$$

$$^{14}_{6}C = {}^{14}_{7}N + e^- + \bar{\nu}$$

$$^{32}_{15}P = {}^{32}_{16}S + e^- + \bar{\nu}$$

As can be seen, during $\beta^-$ decay the mass number (A) remains unchanged and the atomic number (Z) is increased by one. The kinetic energy of the emitted electron is not fixed because the total available energy in the decay (difference between the mass energy of the parent radionuclide and the daughter nuclide) has to be shared between the electron and the antineutrino.

In the above example the difference between the mass energies of $^{3}_{1}H$ and $^{3}_{2}He$, $^{14}_{6}C$ and $^{14}_{7}N$, and $^{32}_{15}P$ and $^{32}_{16}S$ during decay is shared by the electron and the antineutrino. The sharing of the available energy is random and, therefore, the electrons are emitted with varying amounts of kinetic energy (known as the $\beta^-$ spectrum) ranging from 0 to a maximum of $E_\beta$ max. $E_\beta$ max is the total available energy in the case of a particular $\beta^-$ decay. In the above examples $E_\beta$ max for $^{3}H$ decay is 0.018 MeV; for $^{14}C$ decay it is 0.156 MeV; for $^{32}P$ decay it is 1.71 MeV. The probability of an electron emission with a given kinetic energy $E_\beta$, $P(E_\beta)$, varies greatly with

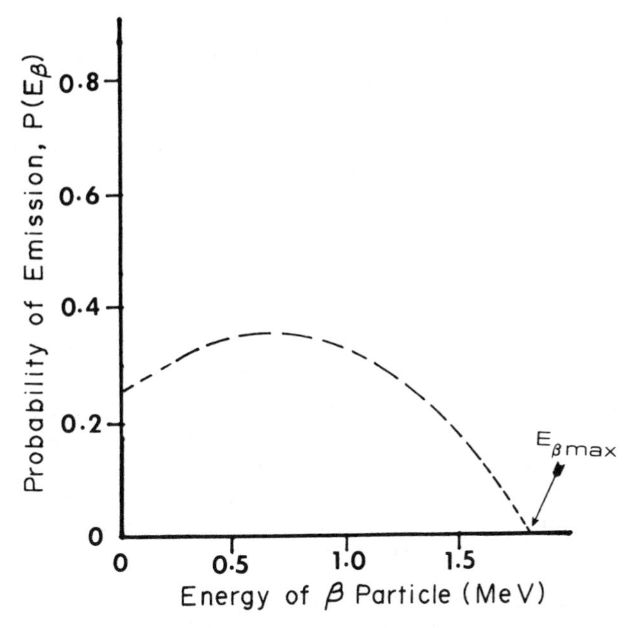


**Fig. 2–2.** A typical $\beta$-ray spectrum. In beta decay, electrons (or positrons) are emitted with varying energies up to a maximum $E_\beta$ max. A beta spectrum gives the probability of emission of an electron with a given energy $E_\beta$.

the kinetic energy $E_\beta$. The variation of $P(E_\beta)$ with $E_\beta$ (*i.e.*, $\beta^-$ spectrum) is shown in Fig. 2–2 for a typical $\beta^-$ decay.

In calculations of radiation dose to a patient with $\beta^-$-emitting radionuclides, one is interested in the average energy, $\overline{E}_\beta$, of the electrons. The computation of $\overline{E}_\beta$ depends on knowledge of the exact shape of $\beta$ spectrum. However, as a rule of thumb, in cases where high accuracy is not desired, $\overline{E}_\beta$ can be obtained by dividing $E_\beta$ max by 3 (*i.e.*, $\overline{E}_\beta = E_\beta \text{ max}/3$). Generally, $E_\beta$ max, and usually $\overline{E}_\beta$, is listed in the table of nuclides.

***$\beta^+$ or Positron Emission:*** In this process, a proton inside the nucleus is converted into a neutron and the excess energy is emitted as a pair of particles, in this case a positron and a neutrino. (Neutrinos and antineutrinos can be considered identical for our purposes.) A positron is an electron with a unit-positive instead of a unit-negative charge. It has the same mass as an electron and interacts with matter in a manner similar to that of an electron. Positron decay can be expressed as:

$${}^{A}_{Z}X = {}_{Z-1}^{\;\;A}Y + e^+ + \nu$$

This equation is consistent with the laws of conservation of mass number and electric charge. However, for the total energy to be conserved, the mass of the nuclide X should be greater than the mass of the Y nuclide by at least 1.02 MeV (2 × mass of an electron). This is because the mass of a nuclide includes in it the mass of the nucleus as well as the electrons associated with it. Therefore, the mass of the X nuclide contains mass of Z electrons, whereas the mass of the Y nuclide contains the mass of only Z − 1 electrons. Also a $e^+$ is created from the nuclear energy of X nuclide. Hence, the mass of the X nuclide has to be greater than 2 × the mass of an electron plus the mass of Y nuclide. Some examples of positron emission are:

$$^{11}_{6}C = ^{11}_{5}B + e^+ + \nu$$

$$^{15}_{8}O = ^{15}_{7}N + e^+ + \nu$$

$$^{18}_{9}F = ^{18}_{8}O + e^+ + \nu$$

Note that during $\beta^+$ emission, the mass number does not change, but the atomic number is decreased by one. In $\beta^+$ emission, too, the energy of the emitted positron varies from 0 to maximum $E_\beta$ max which is known as $\beta^+$ spectrum. The exact computation of the average $\beta^+$ energy, $\overline{E}_\beta{}^+$, in this case, too, is quite involved. As an approximation one can determine $\overline{E}_\beta{}^+$ again by dividing the $E_\beta{}^+$ max by three.

***Electron Capture:*** In this process, a proton inside the nucleus is converted into a neutron by capturing an electron from one of the atomic shells (*e.g.*, K, L, or M). No electron or positron but only a neutrino is emitted. The capture of an electron from the K shell of the atom is known as K capture; capture of an electron from the L shell is known as L capture and so on. Electron capture is one of the few instances (another will be encountered in the case of gamma decay) in which a nucleus interacts directly with the orbital (K, L shells, etc.) electrons of an atom. Once an electron is captured from the K, L or M shell, a vacancy is created in the inner shell of an atom. This vacancy is subsequently filled by electrons from higher shells with simultaneous emission of a characteristic x-ray or Auger electron (see Chapter 1). The probability of a capture from the K shell is generally much higher than that from the L or M shell. Electron capture is expressed as:

$$^{A}_{Z}X + e^- \text{ (orbital)} = {}_{Z-1}^{\ \ A}Y + \mu$$

This is also consistent with the laws of conservation of mass number and electric charge. Common examples of electron capture are:

$$^{51}_{24}Cr + e^+ \text{ (orbital)} = ^{51}_{23}V + \mu$$

$$^{131}_{55}Cs + e^- \text{ (orbital)} = ^{131}_{54}Xe + \mu$$

Note that during electron capture, as in positron emission, the mass number does not change but the atomic number is reduced by one.

During all three processes of beta decay, the mass number A remains unchanged. For this reason beta decay is quite often referred to as an isobaric transition.

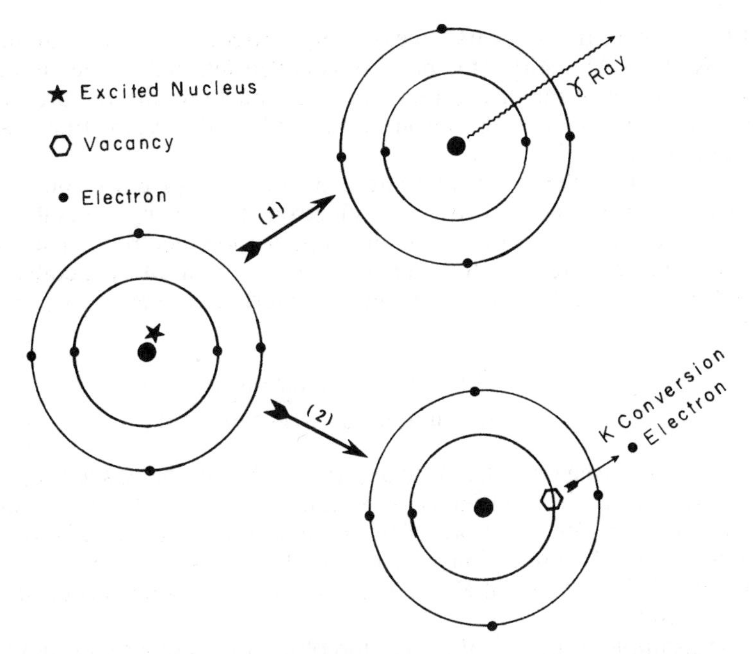


**Fig. 2–3.** Isomeric transition. An excited nucleus in isomeric transition can release its energy in two ways: (1) by the emission of γ-ray (photon), or (2) by the transfer of its energy directly to an orbital electron, generally in the K shell. The electron acquiring the energy is ejected from the atom and is known as the conversion electron. The vacancy thus created in the inner shell of the atom is subsequently filled in the manner discussed in Chapter 1.

## Gamma Decay

It has already been explained that a nucleus can be excited by the absorption of energy to an excited state (isomer). Isomers in general are of very short duration except in cases of metastable states. The decay of an excited state to a lower-energy or ground state is known as isomeric transition (as opposed to isobaric transition in the case of $\beta$ decay) and proceeds through either of two processes: (1) an emission of a high-energy photon, or (2) internal conversion. These are graphically shown in Fig. 2–3.

***High-Energy Photon Emission:*** In this process, the excess energy of an isomer is released in the form of a high-energy photon known as a $\gamma$-ray. A $\gamma$-ray is an electromagnetic radiation with high energy (>100 eV; see Chapter 1). A $\gamma$-ray and an x-ray of the same energy cannot be distinguished from each other because both interact with matter in exactly the same manner. The only difference between the two is that of origin. Energy emitted from the nucleus as a high-energy photon is known as a $\gamma$-ray; energy emitted by an atom (*i.e.*, transition of electrons in atomic shells) as a high-energy photon is called an x-ray. This difference in nomenclature of a high-energy photon is of no practical significance in nuclear medicine.

***Internal Conversion:*** Sometimes a nuclide in an excited state, instead of emitting a $\gamma$-ray, transfers its excess energy directly to an orbital electron (in the K, L or M shell). This process is called internal conversion and is another example of the direct interaction of a nucleus with the orbital electrons revolving around it. (Electron capture in beta decay, the first example, is described on p. 17.)

Internal conversion is an alternate process to $\gamma$-ray emission. A single nucleus will emit either a $\gamma$-ray or an electron. However, in a collection of nuclei, some will emit $\gamma$-rays and some conversion electrons. The ratio of the number of electrons to the number of $\gamma$-rays emitted by a collection of excited nuclei is called the coefficient of internal conversion (ic) for that excited state:

$$\text{ic} = \frac{\text{total number of electrons emitted}}{\text{total number of } \gamma\text{-rays emitted}}$$

When considering only the electrons knocked from the K shell, the above ratio is known as the coefficient of K conversion, $\text{ic}_K$. When considering only the electrons knocked out of the L shell, the above ratio is known as the coefficient of L conversion, $\text{ic}_L$. The total conversion coefficient $\alpha$, then, is the sum of the individual K, L, M conversion coefficients (*i.e.*, $\text{ic} = \text{ic}_K + \text{ic}_L + \text{ic}_M$).

As an example, if $\text{ic} = \frac{1}{5}$, the emission of a $\gamma$-ray is five times more likely than the emission of an electron by internal conversion or out of six decays, five will be through $\gamma$-ray or photon emission and only one will be through the conversion process. Therefore if there are 100 decays of excited nuclei, then in $\frac{5}{6} \times 100 \simeq 83$ decays a $\gamma$-ray will be emitted, and in the remaining 17 decays a conversion electron will be emitted. If the internal conversion coefficient, ic, is 2, then the emission of an electron is twice as likely as the emission of a $\gamma$-ray, or out of three decays, one will be through $\gamma$-ray emission and two will be through conversion electron emission. Therefore if there are 100 decays of the excited nuclei, then the number of $\gamma$-rays emitted will be $\frac{1}{3} \times 100 \simeq 33$, whereas the number of conversion electrons emitted will be 67. When I discuss later the ideal properties of a radionuclide for nuclear medicine (Chap. 5), it will be seen that radionuclides that do not emit particulate radiation are desirable. Since even in gamma decay, because of internal conversion, electrons can be emitted, it follows that the coefficient of internal conversion should be as small as possible.

The probability of K conversion is generally quite high compared to L and M conversion because of the closeness of the K shell electrons to the nucleus compared to the L and M shell electrons. Also, the probability of internal conversion is higher if the excited state is long-lived (metastable state) and the energy of the excited state is low (<100 keV).

Since in internal conversion an electron is emitted from the inner shell of an atom (K, L or M), a vacancy is created in that shell. This vacancy is subsequently filled by electrons from higher shells leading to the emission of an x-ray or Auger electron. The process of vacancy filling in atomic shells is the same as that following K capture in beta decay. As the electrons in each shell (K, L, . . . ) of the atom are bound with a certain amount of energy known as binding energy, the conversion electron carries a kinetic energy which is the difference of the energy of the excited state (nucleus) and the binding energy of the electron in a given shell. For example, $^{113m}$In, a metastable state with an energy of 393 keV, emits a $\gamma$-ray of 393 keV, but emits a K conversion electron with a kinetic energy of 393 − 29.7 (B.E. of K electrons in In atom) = 363.3 keV.

## DECAY SCHEMES

In this discussion of alpha, beta and gamma decay, various examples of radioactive processes have been presented. How does one know that a given radionuclide will decay by the emission of an $\alpha$ particle, $\beta$-ray, or $\gamma$-ray, or by a combination of two or more? Actually, thee are no theoretical laws which can supply this information. Such data are experimentally determined for each radionuclide and then tabulated in a pictorial form known as a decay scheme. A decay scheme is a collection of experimental information regarding the modes and frequency of decay, process of decay, energy of different radiation emitted, half-life and other information of interest, for a given radionuclide.

In one commonly employed approach, decay schemes of isobars (or $\beta$ transitions) are graphically represented on one page, with radionuclides arranged in order of increasing atomic number from left to right. In this representation, all $\beta^-$-decaying radionuclides decay to the nuclide on their right (↘) and all the K-capture or $e^+$-decaying radionuclides to their left (↙ or ⤦). A broken arrow is the indication for positron decay. It signifies that the mass of the parent nuclide has to be greater than the mass of the daughter nuclide by at least 1.02 MeV. The excited states of the nuclides are shown as horizontal bars above the ground state. The energy of these states increases from the bottom to the top of the page. Isomeric transitions between two states are shown by vertical lines connecting the two states.

In this instance, as opposed to the energy of electrons in an atom (p. 4), the zero of the energy scale is, by convention, chosen to be the ground state of a nuclide. It is as if the K shell of an atom has been arbitrarily chosen to have zero energy. According to this convention, then all other shells of the atom (L, M, etc.) will have positive energy. In nuclear convention then, the energy of the electrons in the K, L, M shells of a sodium atom will, instead of −1072, −63 and 1 eV, be 0, 1009 and 1071 eV respectively.

A decay scheme is shown in Fig. 2–4 for $^{99}$Mo, the parent of $^{99m}$Tc which has revolutionized the field of nuclear medicine. Briefly, the decay scheme of $^{99}$Mo shows that $^{99}$Mo decays 80% of the time to an excited state of $^{99}$Tc, which has an energy of 0.142 MeV (a metastable state because of its relatively long life), by emission of a $\beta^-$ particle with $E_\beta$ max = 1.23 MeV. For 18.5% of the time, the decay proceeds to another excited state (isomer) of $^{99}$Tc, which has an energy of 0.922 MeV, by emission of a $\beta^-$ particle with $E_\beta$ max = 0.45 MeV. For the remaining 1.4% of the time, the decay takes place to other excited states of $^{99}$Tc. The excited state of $^{99}$Tc at 0.922 MeV is short-lived and quickly ($<10^{-9}$ sec) decays to the ground state or to lower-energy excited states by emission of $\gamma$-rays or conversion electrons.

Table 2–1 lists the number of radiations eventually emitted in the decay

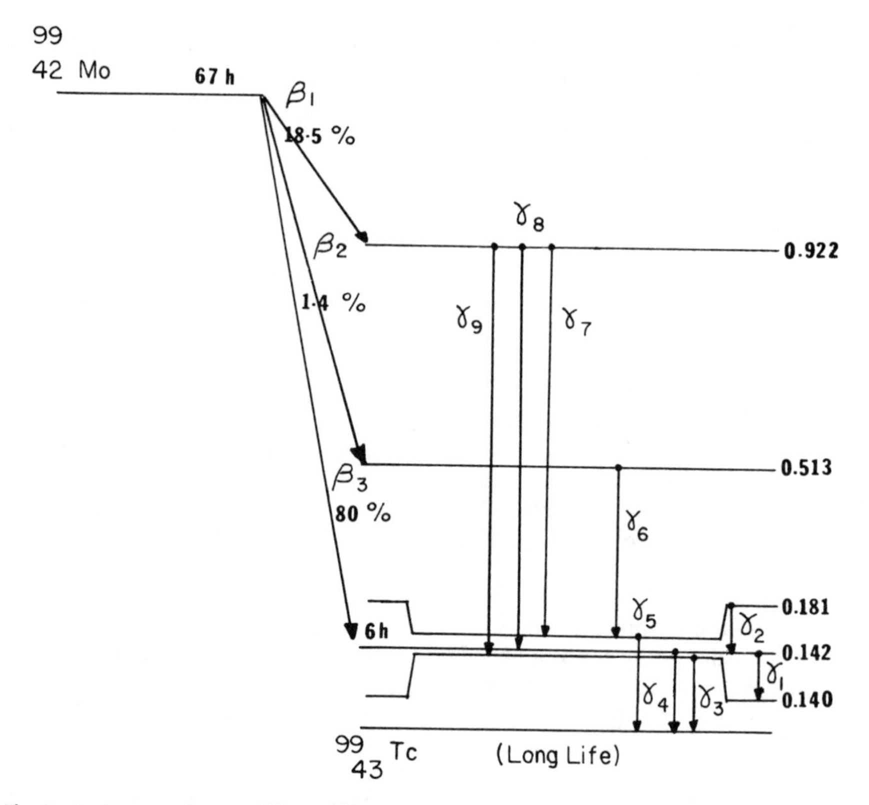


**Fig. 2–4.** Decay scheme of $^{99}$Mo. $^{99}$Mo decays via $\beta$-emission to the 0.142, 0.513 and 0.922 MeV excited states of $^{99}$Tc. These excited states then decay to the ground state by the emission of a number of $\gamma$-rays shown above. Less likely transitions have been omitted from the decay scheme. Except for the 0.142 MeV state, which has a half-life of 6 hr, other excited states are short-lived.

of $^{99}$Mo ($\beta$-rays, $\gamma$-rays, characteristic x-rays, conversion electrons, and Auger electrons) with their respective energies, $\overline{E}_i$, and the frequency of emission $n_i$ ($n_i$ is defined as the probability of emission of radiation i with energy $E_i$ per decay of a radionuclide). The information given in Table 2–1, however, cannot be obtained from Fig. 2–4 alone because additional data and complex computations are required. The *Journal of Nuclear Medicine* (Suppl. 10, 1975) has published such information for a large number of radionuclides.

Two more examples of decay schemes, $^{99m}$Tc and $^{125}$I, are also given here. Appendix A gives in tabular form radiations emitted by a number of other radionuclides commonly used in nuclear medicine. Only those radiations which are emitted more than 1.0% of the time in the decay are included in these tabulations which are primarily derived from the above source. For radionuclides not given here or in Appendix A, the reader should consult the above reference. Although information regarding the decay of $^{99m}$Tc can be derived from the decay scheme of $^{99}$Mo, the decay scheme of $^{99m}$Tc is

**Table 2–1. Radiations Emitted in the Decay of $^{99}Mo$**

| Number | Radiation (i) | Frequency of Emission ($n_i$) | Mean Energy (MeV) ($\bar{E}_i$) |
|---|---|---|---|
| 1 | $\beta$ 1 | 0.185 | 0.140 |
| 2 | $\beta$ 2 | 0.014 | 0.298 |
| 3 | $\beta$ 3 | 0.797 | 0.452 |
| 4 | $\gamma$ 1 | — | — |
| 5 | M Conversion Electron | 0.851 | 0.002 |
| 6 | $\gamma$ 2 | 0.130 | 0.041 |
| 7 | K Conversion Electron | 0.043 | 0.019 |
| 8 | $\gamma$ 3 | 0.815 | 0.140 |
| 9 | K Conversion Electron | 0.085 | 0.120 |
| 10 | L Conversion Electron | 0.011 | 0.138 |
| 11 | $\gamma$ 4 | — | — |
| 12 | $\gamma$ 5 | 0.066 | 0.181 |
| 13 | $\gamma$ 6 | 0.014 | 0.366 |
| 14 | $\gamma$ 7 | 0.137 | 0.740 |
| 15 | $\gamma$ 8 | 0.048 | 0.778 |
| 16 | X-Rays—K($\alpha$) | 0.094 | 0.018 |
| 17 | X-Rays—K($\beta$) | 0.017 | 0.021 |
| 18 | KLL Auger Electron | 0.022 | 0.015 |
| 19 | KLX Auger Electron | 0.01 | 0.018 |
| 20 | LMM Auger Electron | 1.53 | 0.002 |
| 21 | MXY Auger Electron | 1.20 | 0.001 |

shown separately in Fig. 2–5 because of its great importance in nuclear medicine. Table 2–2 lists the relevant data for $^{99m}Tc$.

Decay scheme for $^{125}I$, a radionuclide which is widely used in radioimmunoassays, is shown in Fig. 2–6. The number of $\gamma$-rays, x-rays, conversion

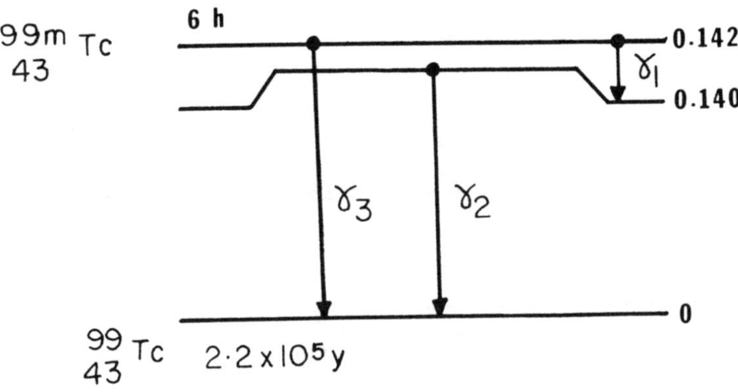


**Fig. 2–5.** Decay scheme of $^{99m}Tc$: the 6-hr half-life isomeric state of $^{99}Tc$ at 0.142 MeV primarily (99% probability) decays to another isomeric state at 0.140 MeV. It has a very short half-life ($10^{-9}$ sec) and, therefore, it decays almost instantaneously to the ground state by the emission of a 0.140 MeV $\gamma$-ray or a corresponding conversion electron. The ground state or the nuclide $^{99}Tc$ itself is a radionuclide. However, its half-life is so long that for all practical purposes it may be considered stable.

**Table 2–2. Radiations Emitted in the Decay of $^{99m}Tc$**

| *Number* | *Radiation (i)* | *Frequency of Emission* ($n_i$) | *Mean Energy (MeV)* ($\bar{E}_i$) |
|---|---|---|---|
| 1 | $\gamma$ 1 (Conversion Electron) | 0.986 | 0.002 |
| 2 | $\gamma$ 2 (Photon) | 0.883 | 0.140 |
| 3 | K Conversion Electron | 0.088 | 0.119 |
| 4 | L Conversion Electron | 0.011 | 0.138 |
| 5 | M Conversion Electron | 0.004 | 0.140 |
| 6 | $\gamma$ 3 (Conversion Electron) | 0.01 | 0.122 |
| 7 | K ($\alpha$) x-Ray | 0.064 | 0.018 |
| 8 | K ($\beta$) x-Ray | 0.012 | 0.021 |
| 9 | KLL Auger Electron | 0.015 | 0.015 |
| 10 | LMM Auger Electron | 0.106 | 0.002 |
| 11 | MXY Auger Electron | 1.23 | 0.0004 |

electrons, etc., and their respective energies and frequency of emission in the decay of $^{125}I$ are given in Table 2–3.

It is important to remember that, in the decay of $^{99}Mo$ and $^{125}I$, the $\gamma$-rays and x-rays emitted are not from $^{99}Mo$ and $^{125}I$ nuclides, but from $^{99}Tc$ and $^{125}Te$ nuclides, respectively, even though in laboratory jargon these are re-

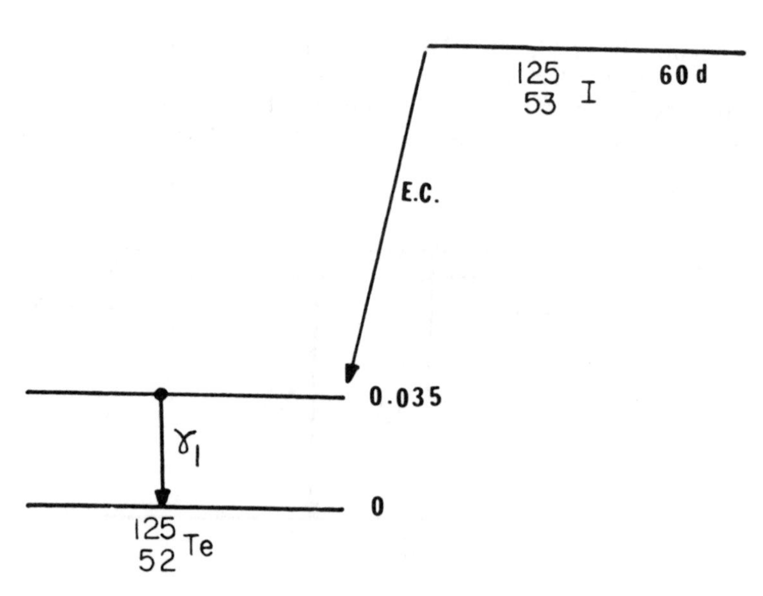


**Fig. 2–6.** Decay scheme of $^{125}I$: $^{125}I$ decays through electron capture to an isomeric state of $^{125}Te$ at 0.035 MeV which subsequently decays to the ground state by the emission of a $\gamma$-ray (7% probability) or a conversion electron (93% probability). Since in electron capture as well as internal conversion a vacancy is created in the K shell, K x-rays, of $^{125}Te$ (29.8 keV) are also emitted in the decay of $^{125}I$.

**Table 2–3. Radiations Emitted in the Decay of $^{125}I$**

| Number | Radiation (i) | Frequency of Emission ($N_i$) | Mean Energy (MeV) ($\bar{E}_i$) |
|---|---|---|---|
| 1 | $\gamma$ 1 | 0.068 | 0.035 |
| 2 | K Conversion Electron | 0.746 | 0.004 |
| 3 | L Conversion Electron | 0.107 | 0.031 |
| 4 | M Conversion Electron | 0.080 | 0.035 |
| 5 | X-Ray—K($\alpha$) | 1.176 | 0.027 |
| 6 | X-Ray—K($\beta$) | 0.240 | 0.031 |
| 7 | X-Ray—L | 0.215 | 0.004 |
| 8 | KLL Auger Electron | 0.137 | 0.023 |
| 9 | KLX Auger Electron | 0.058 | 0.026 |
| 10 | KXY Auger Electron | 0.01 | 0.030 |
| 11 | LMM Auger Electron | 1.49 | 0.003 |
| 12 | MXY Auger Electron | 3.59 | 0.001 |

ferred to as iodine 125 $\gamma$-rays or molybdenum 99 $\gamma$-rays. This is a misnomer. The decay actually takes place in the following series:

$$^{99}\text{Mo} \xrightarrow{\beta-} {}^{99}\text{Tc}^* \xrightarrow{\gamma} {}^{99}\text{Tc}$$

$$^{125}\text{I} \xrightarrow{\text{K Capture}} {}^{125}\text{Te}^* \xrightarrow{\gamma} {}^{125}\text{Te}$$

## PROBLEMS

1. Separate the following nuclides into pairs of isotopes, isobars, isotones or isomers:

$$^{3}_{1}\text{H},\ ^{4}_{2}\text{He},\ ^{3}_{2}\text{He},\ ^{12}_{6}\text{C},\ ^{12}_{7}\text{N},\ ^{14}_{6}\text{C},\ ^{99}_{43}\text{Tc},\ ^{99}_{42}\text{Mo},\ ^{99m}_{43}\text{Tc},\ ^{100}_{44}\text{Ru}.$$

2. Identify the following decay processes in which: (a) the mass number is unchanged, (b) the atomic number decreases by one, (c) the atomic number increases by one, (d) the mass number decreases by four, and (e) the mass and atomic numbers remain constant.
3. The $E_{\beta\ max}$ for $^{3}H$ and $^{32}P$ are 0.0186 and 1.710 MeV, respectively. What are the average energies of β emission in these decays?
4. In electron capture, an electron is removed from the K or L shell of an atom. Is the daughter atom in the ground state or excited state? Is it ionized?
5. Can a radionuclide decay with all three ($\alpha$, $\beta$, and $\gamma$) processes?
6. Find the following radiations in the decay tables given in this chapter or in Appendix A: (a) x-ray with the highest energy in the decay of $^{99m}Tc$; (b) $\gamma$ rays in the energy range of 100 to 300 keV in the decay of $^{111}In$; (c) $\beta$ ray with the least frequency of emission, $\beta$ ray with the most energy of emission and $\gamma$ rays in the energy range of 700 to 800 keV in the decay of $^{99}Mo$; and (d) % emission and the energies of x-rays in the decay of $^{201}Tl$.

# 3

# Radioactivity—Law of Decay, Half-Life and Statistics

Radionuclides are unstable and decay to other nuclides by the processes discussed previously. At what rates do radionuclides decay? Do all radionuclides decay at the same rate? How many radioatoms (*i.e.*, an atom whose nucleus is radioactive) remain after a given period of time? The answers to these and related questions constitute the primary subject matter of this chapter.

## RADIOACTIVITY—DEFINITION AND UNITS

The number of disintegrations per unit of time (decay rate) is called the radioactivity or simply the activity of a given sample of a radionuclide. For example, if we have a collection of 1,000 atoms of a radionuclide at a given time and 50 of these disintegrate in the next 5 seconds, the activity of this sample is $\frac{50}{5} = 10$ disintegrations per second (dis/sec). In algebraic terms, if out of a number $N_t$ of radioatoms at a time t, a number of radioatoms, $dN_t$, decays in a small time interval dt, the radioactivity, $R_t$, of the sample is calculated by dividing the number of radioatoms decayed ($dN_t$) by the interval (dt) during which the decay took place. In other words,

$$R_t = \frac{-dN_t}{dt} \qquad [1]$$

In this equation the minus sign denotes that the radioatoms are decreasing in number.

The unit of radioactivity is the curie (Ci) in the older system of units (CGS) and the becquerel (Bq) in the newer system of units (SI). SI units are now in the process of replacing the older system of units in scientific work. Unfortunately, in clinical nuclear medicine, old units are still popular. Therefore I have decided to keep the old units. However, to encourage the use and familiarity of new units, these are given in parentheses for the important data, results or problems. The major differences between the two systems of units are given in Appendix B.

A curie is defined as the radioactivity of a sample which is disintegrating at a rate of $3.7 \times 10^{10}$ disintegrations/second, whereas a becquerel is defined as the radioactivity of a sample which is disintegrating at a rate of 1 disintegration/second. Since the curie is a large unit, the smaller units derived from it are often used. On the other hand, the becquerel is a small unit, and therefore larger units derived from it are used. Some of these derived units for the curie as well as the becquerel and their interrelationships are given below:

$$\text{Curie (Ci)} = 3.7 \times 10^{10} \text{ dis/sec} = 37 \text{ GBq}$$

$$\text{Millicurie (mCi)} = 10^{-3} \times \text{Ci} = 3.7 \times 10^{7} \text{ dis/sec} = 37 \text{ MBq}$$

$$\text{Microcurie } (\mu\text{Ci}) = 10^{-6} \times \text{Ci} = 3.7 \times 10^{4} \text{ dis/sec} = 37 \text{ KBq}$$

$$\text{Nanocurie (nCi)} = 10^{-9} \times \text{Ci} = 3.7 \times 10 \text{ dis/sec} = 37 \text{ Bq}$$

$$\text{Picocurie (pCi)} = 10^{-12} \times \text{Ci} = 3.7 \times 10^{-2} \text{ dis/sec.}$$

and

$$1\text{Bq} = 1 \text{ dis/sec} = 27.03 \times 10^{-12} \text{ Ci} = 27.03 \text{ pCi}$$

$$\text{Kilobecquerel (KBq)} = 10^{3} \text{ Bq} = 27.03 \text{ nCi}$$

$$\text{Megabecquerel (MBq)} = 10^{6} \text{ Bq} = 27.03\ \mu\text{Ci}$$

$$\text{Gigabecquerel (GBq)} = 10^{9} \text{ Bq} = 27.03 \text{ mCi}$$

## LAW OF DECAY

When a sample of a radionuclide is observed for a long period of time, it is found that the radioactivity ($R_t$) of a sample at a given time (t) depends on: (1) The number of the radioatoms ($N_t$) present at that time and, (2) a constant ($\lambda$) which is characteristic of a given radionuclide and is different for various radionuclides.

The constant $\lambda$, also known as the decay constant, is defined as the probability of decay per unit of time for a single radioatom. Therefore, if there are $N_t$ radioatoms present at time t, then the number of nuclei disintegrating per unit time ($R_t$) at time (t) will be simply the product of $\lambda$ and $N_t$ or

$$R_t = \lambda \cdot N_t \qquad [2]$$

From this equation it can be seen that for the different radionuclides (*i.e.*, different $\lambda$), the number of radioatoms ($N_t$) that must be present to produce the same amount of radioactivity ($R_t$) varies. For example, the number of radioatoms present in a sample of $^{99m}$Tc, which has a radioactivity of 1 mCi (37 MBq), is not the same as the number of radioatoms present in a sample of $^{131}$I, which has a radioactivity of 1 mCi (37 MBq). This is true because the decay constant for $^{99m}$Tc ($\lambda = 3.2 \times 10^{-5}$/sec) is different from the decay constant for $^{131}$I ($\lambda = 10^{-6}$/sec). The number of radioatoms of $^{99m}$Tc or of $^{131}$I which must be present in a sample to produce a radioactivity of 1 mCi can easily be determined from equation (2). For $^{99m}$Tc,

$$R_t = 1 \text{ mCi (37 MBq)} = 3.7 \times 10^7 \text{ dis/sec and } \lambda(^{99m}\text{Tc}) = 3.2 \times 10^{-5}/\text{sec}$$

Therefore, from equation (2),

$$N_t\,(^{99m}\text{Tc}) = \frac{R_t}{\lambda} = \frac{3.7 \times 10^7}{3.2 \times 10^{-5}} = 1.15 \times 10^{12} \text{ radioatoms}$$

For $^{131}$I,

$$R_t = 1 \text{ mCi (37 MBq) and } \lambda\,(^{131}\text{I}) = 10^{-6}/\text{sec}$$

Therefore, from equation (2),

$$N_t\,(^{131}\text{I}) = \frac{3.7 \times 10^7}{10^{-6}} = 3.7 \times 10^{13} = 32 \times N_t\,(^{99m}\text{Tc}) \text{ radioatoms}$$

(*i.e.*, for the same amount of radioactivity there are approximately 32 times more radioatoms present in a sample of $^{131}$I than in a sample of $^{99m}$Tc).

**Calculation of the Mass of a Radioactive Sample**

Once we know the number of radioatoms $N_t$ present in a 1-mCi sample of a radionuclide of mass number A, it is simple to calculate the mass of the radionuclide present in that sample. Here we assume that the radioactive sample contains only the radionuclide under consideration and none of its isotopes.

From equation 2, for a 1-mCi sample, $N_t = \dfrac{3.7 \times 10^7}{\lambda}$. From Avogadro's hypothesis, A grams of this radionuclide contain a total of $6 \times 10^{23}$ radioatoms. Therefore,

$$\text{the mass of 1 radioatom} = \frac{A}{6 \times 10^{23}} \text{ gm}$$

$$\text{or, mass, M, of } N_t \text{ radioatoms} = \frac{N_t \cdot A}{6 \times 10^{23}} \text{ gm}$$

$$\text{or, substituting value of } N_t, = \frac{3.7 \times 10^7 \times A}{6 \times 10^{23} \times \lambda} \text{ gm/mCi}$$

$$\text{or, mass, M} = 6 \times 10^{-17} \cdot \frac{A}{\lambda} \text{ gm/mCi}$$

$$\left[1.67 \times 10^{-8} \frac{A}{\lambda} \text{ gm/MBq}\right]$$

*Example:*

What is the mass of a 1-mCi sample of $^{99m}$Tc? The mass number for this radionuclide is 99 and the decay constant is $3.2 \times 10^{-5}$/sec. Therefore,

$$M = 6.10^{-17} \times \frac{99}{3.2 \times 10^{-5}} = 1.8 \times 10^{-10} \text{ gm}$$

## Specific Activity

In the above calculations the mass of a 1-mCi sample was determined. In many instances, however, one is interested in the reciprocal of the above, *i.e.*, the amount of radioactivity per unit mass of the radionuclide of interest. This, the radioactivity per unit mass of a radionuclide, is known as the specific activity of a radioactive sample. It is generally expressed in units such as mCi/mg (MBq/mg) or disintegration/min/mg. In the above example, the specific activity of the $^{99m}$Tc sample is simply the inverse of M or equal to $1/(1.8 \times 10^{-10})$ mCi/gm or equal to $5.5 \times 10^{9}$ mCi/gm.

In this example, it was assumed that all the atoms of technetium were radioactive and there were no stable or longer lived technetium atoms present in the sample. Such a sample which contains only the radionuclide of interest and no other isotope or long lived radioisotope of that radionuclide is generally called a "carrier free" sample. Therefore, the above calculation of the specific activity applies only to a carrier free sample. The specific activity of a sample with carrier (which contains stable or longer lived isotopes of the radionuclide of interest) will always be less than that of a carrier free sample and will depend upon the amount of the carrier in the sample.

When a radionuclide is not present in its elemental form, but is a part of a molecule, the concept of specific activity can be generalized to the mass of the molecule rather than the element. For instance, in the above sample, if the technetium radioactivity is present as pertechnetate ($^{99m}TcO_4^-$ rather than as a simple element ($^{99m}$Tc), one can speak of the specific radioactivity of the sample per unit mass of the pertechnetate. Again, if the sample is carrier free, (*i.e.*, no molecules of pertechnetate exist in the sample which do not contain $^{99m}$Tc as its part or $^{99}TcO_4^-$ molecules are absent), the specific activity of a pertechnetate sample can be calculated as above by simply taking the inverse of M and substituting the molecular weight of $TcO_4^-$ for A in the above formula. Since the molecular weight of $TcO_4^-$ is about 163, the specific activity of 1-mCi sample of $^{99m}TcO_4^-$ is $= \frac{1}{M} = \frac{\lambda}{A \times 6 \times 10^{-17}} = \frac{3.2 \times 10^{-5}}{163 \times 6 \times 10^{-17}} = 3.4 \times 10^{9}$ mCi/gm ($1.18 \times 10^{11}$ MBq/gm). Of course, when the sample is not carrier free, the specific activity of the sample will be lower than that of a carrier free sample and will depend on the amount of carrier present. In nuclear medicine, the knowledge of the specific activity of a sample is important from several considerations: (1) the toxicity of a chemical or drug is always dependent on the amount of the chemical or drug administered to the patient. Therefore, it is important to know the amount of a particular chemical or drug in the radioactive sample.

(2) The distribution of a radiochemical in a biological system many times depends on the amount of carrier present in the sample. (3) The labeling efficiency of a radionuclide to a chemical is also sometimes dependent on the specific activity of the radionuclide.

### The Exponential Law of Decay

If we combine equations (1) and (2), then $-\frac{dN_t}{dt} = \lambda N_t$. This is a differential equation whose solution is given by the following expression:

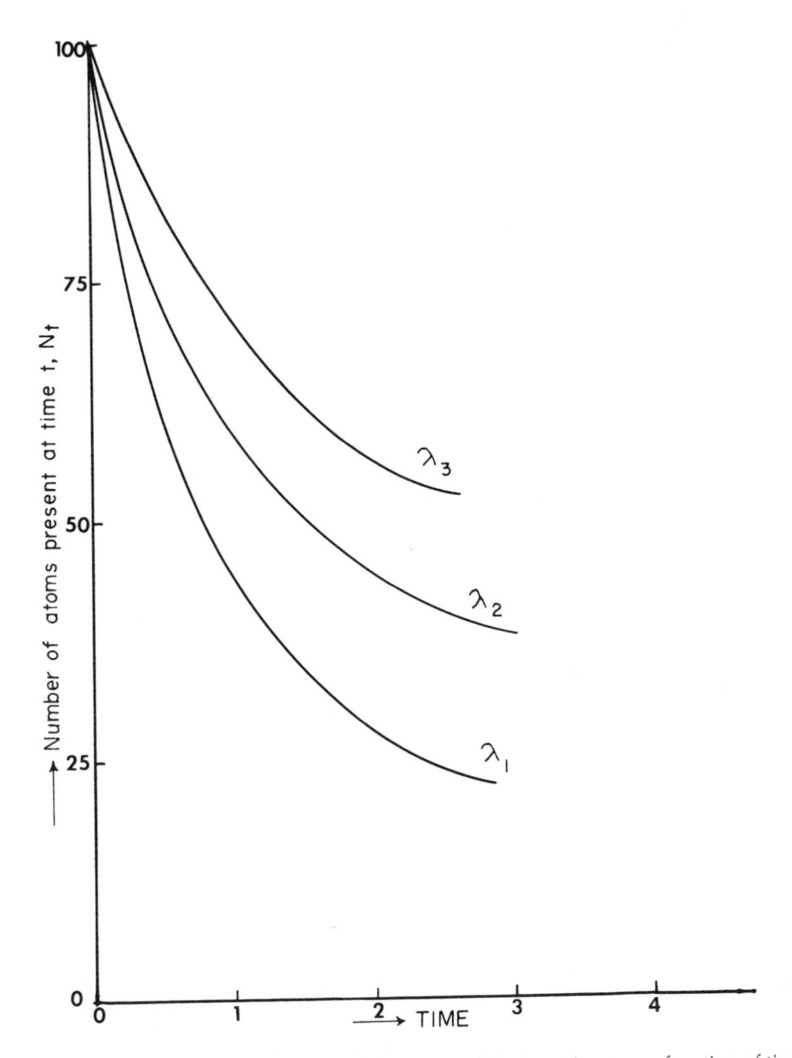


**Fig. 3–1.** Linear plot of the number of radioatoms remaining at a time t as a function of time for three radionuclides with different decay constants, $\lambda_1$, $\lambda_2$, $\lambda_3$.

$$N_t = N_0 e^{-(\lambda \cdot t)} \qquad [3]$$

This equation relates the number of radioatoms $N_t$ remaining at time t when we initially started (at $t = 0$) with a number $N_0$, and is known as the exponential law of decay. If we plot this expression as a linear graph, curves similar to those shown in Fig. 3–1 result for different radionuclides (different $\lambda$). When one plots the same expression on semilog graph paper, as in Fig. 3–2, one obtains straight lines of different slopes (steepness) for different radionuclides.

The exponential law of decay is also true for the radioactivity $R_t$—that

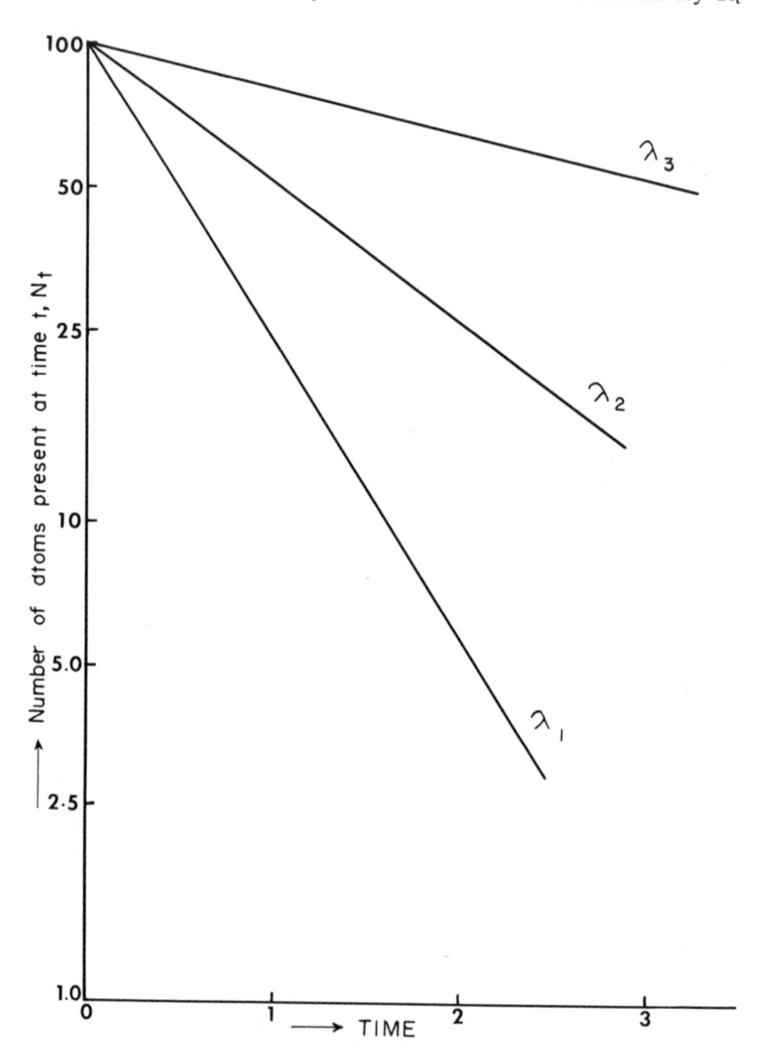


**Fig. 3–2.** The same data as in Fig. 3–1, plotted here on a semi-logarithmic graph. Plot of exponential function on semi-log graph paper results in a straight line. The slope of the straight line is determined by the decay constant or the half-life of the radionuclide.

is, $R_t$ is also related to the original radioactivity $R_0$ by an expression similar to equation 3. This can be seen easily by substituting equation 3 on the right-hand side of equation 2:

$$R_t = \lambda \cdot N_t$$
$$= \lambda \cdot N_0 \cdot e^{-(\lambda \cdot t)} \qquad [4]$$
$$\text{or } R_t = R_0 \cdot e^{-(\lambda \cdot t)}$$

where $R_0 = \lambda \cdot N_0$ is the radioactivity at time zero. If one plots activity $R_t$ as a function of time, curves similar to those shown in Figs. 3–1 and 3–2 result.

## Half-Life

How does one determine the decay constant $\lambda$? Experimentally it is more convenient to measure another parameter, $T_{\frac{1}{2}}$, known as the half-life. This is defined as the time interval for a given number of nuclei (or their radioactivity) to decay to one-half of the original value. For example, if initially there are 10,000 nuclei in a radionuclide and 5,000 of these decay in 5 days, then the half-life of this radionuclide is 5 days. Therefore, by definition of the half-life, when $N_t = \frac{N_0}{2}$, then $t = T_{\frac{1}{2}}$.

If we substitute these values in equation 3 we obtain

$$\frac{N_0}{2} = N_0 e^{-(\lambda \cdot T_{\frac{1}{2}})}$$

or

$$\tfrac{1}{2} = e^{-(\lambda \cdot T_{\frac{1}{2}})}$$

Since $e^{-0.693}$ is equal to $\frac{1}{2}$ (see Appendix C), it follows that

$$\lambda \cdot T_{\frac{1}{2}} = 0.693 \qquad [5]$$

Equation 5 relates the decay constant to the half-life of a radionuclide. From this, if one knows the half-life of the radionuclide, one can calculate the decay constant or vice versa.

*Example:*
Half-life of $^{131}I$ = 8 days. Determine the decay constant for $^{131}I$.

$$T_{\frac{1}{2}} = 8 \text{ days} = 8 \times 24 \times 60 \times 60 \text{ sec}$$
$$= 0.691 \times 10^6 \text{ sec}$$

From equation 5,

$$0.691 \times 10^6 \times \lambda = 0.693 \text{ or } \lambda = 10^{-6}/\text{sec}$$

## Problems on Radioactive Decay

Equations 1 to 5 are the basic equations necessary for solving routine problems in nuclear medicine. In general, to solve such problems use of exponential tables is necessary (Appendix C). However, in special cases, where multiples of half-life are involved, equation 3 can be written in a simpler form by substituting the value of λ from equation 5.

$$N_t = N_0 e^{-\frac{(0.693t)}{T_{\frac{1}{2}}}}$$

$$= N_0(e^{-0.693})^{\frac{t}{T_{\frac{1}{2}}}} \qquad [6a]$$

$$= N_0(\tfrac{1}{2})^{\frac{t}{T_{\frac{1}{2}}}}$$

When t is a multiple of $T_{\frac{1}{2}}$, it is easier to make calculations using the above expression than to use exponential tables. For example, if one is interested in finding the number of radioatoms remaining after an interval $t = 3 \times$ the half-life of a radionuclide, the above expression becomes

$$N_t = N_0 \cdot (\tfrac{1}{2})^{\frac{3 \times T_{\frac{1}{2}}}{T_{\frac{1}{2}}}} = N_0(\tfrac{1}{2})^3$$

$$= N_0/8$$

or, one-eighth of the original number. A similar simplification occurs for the activity:

$$R_t = R_0(\tfrac{1}{2})^{\frac{t}{T_{\frac{1}{2}}}} \qquad [6b]$$

*Example 1:*

A radioactive sample of $^{99m}$Tc contains 10-mCi (370 MBq) radioactivity at 9:00 A.M. What will be the radioactivity of this sample at 3:00 P.M. on the same day? (The half-life of $^{99m}$Tc is 6 hr.) In this example,

$$R_0 = 10 \text{ mCi (370 MBq)}$$

$$T_{\frac{1}{2}} = 6 \text{ hr}$$

$$t = 3{:}00 \text{ P.M.} - 9{:}00 \text{ A.M.} = 6 \text{ hr}$$

Now, using equation [6b]

$$R_t = 10 \cdot (\tfrac{1}{2})^{6/6} = 10/2 = 5 \text{ mCi (185 MBq)}$$

*Example 2:*

A preparation of $^{99m}$Tc was calibrated at 7 A.M. and contained 15 mCi/ml (555 MBq/ml) of radioactivity at that time. If the desired amount of radioactivity to be administered to the patient is 15 mCi (555 MBq), determine the volume of the preparation which will have to be injected at 10 A.M.

$$R_0 = 15 \text{ mCi/ml}$$

$$T_{\frac{1}{2}} = 6 \text{ hours or } \lambda = \frac{0.693}{6} \text{/hr}$$

$$t = 10 \text{ A.M.}-7 \text{ A.M.} = 3 \text{ hours}$$

Using equations 4 and 5

$$R_t \text{ (at 10 A.M.)} = 15 \cdot e^{\frac{-0.693 \times 3}{6}} = 15e^{-0.346}$$

$$= 15 \times 0.70 \text{ (from Appendix C)}$$

$$= 10.5 \text{ mCi/ml (388.5 MBq/ml)}$$

Amount desired = 15 mCi (555 MBq)

$$\therefore \text{Volume needed} = \frac{15 \text{ mCi}}{10.5 \text{ mCi/ml}} = 1.43 \text{ ml}$$

*Example 3:*

An iodine-123 capsule is calibrated to contain 100 $\mu$Ci (3.7 MBq) at noon. However, the capsule is given to a patient early in the morning at 9 A.M. Find the amount of the radioactivity administered to the patient. (Half-life of $^{123}$I = 13 hours.)

Let us assume that the radioactivity at 9 A.M. is $R_0$. Then, the radioactivity at noon (*i.e.*, 3 hours later) $R_t$ = 100 $\mu$Ci. Using equations 4 and 5 we get

$$100 = R_0 \cdot e^{\frac{-0.693 \times 3}{13}} = R_0 \cdot e^{-0.16}$$

$$= R_0 \times 0.85 \text{ (from Appendix C)}$$

$$R_0 = \frac{100}{0.85} = 117.6 \ \mu\text{Ci (4.3 MBq)}$$

Quite often, the term $e^{-\lambda \cdot t}$ is tabulated for various intervals of time for a given radionuclide. These are known as decay factors. In example 3, the decay factor for $^{123}$I for 3 hour decay is 0.85. If the table of decay factors is available, then the relationship of $R_t$ and $R_0$ becomes simplified as $R_t = R_0 \times$ Decay Factor, where Decay Factor = $e^{-\lambda \cdot t}$. Table 3–1 lists some of the decay factors for $^{99m}$Tc.

For practical considerations, a simple fact to remember is that the radioactivity remaining after 10 half-lives of a radionuclide is about one-thousandth of the original radioactivity (*i.e.*, millicurie amounts are reduced to

**Table 3–1. Decay Factors for $^{99m}Tc$**

| *Elapsed Time (h)* | *Decay Factor* | *Elapsed Time (h)* | *Decay Factor* |
|---|---|---|---|
| 0.5 | 0.94 | 3.5 | 0.67 |
| 1.0 | 0.89 | 4.0 | 0.63 |
| 1.5 | 0.84 | 4.5 | 0.59 |
| 2.0 | 0.79 | 5.0 | 0.56 |
| 2.5 | 0.74 | 5.5 | 0.53 |
| 3.0 | 0.71 | 6.0 | 0.50 |

microcurie amounts or MBq amounts are reduced to kBq amounts). This is so because $(\frac{1}{2})^{10} \sim \frac{1}{1000}$.

### Average Life ($T_{av}$)

Occasionally we encounter the use of another term known as average life of a radionuclide. It is related to the half-life of the radionuclide or the decay constant $\lambda$ as follows:

$$T_{av} = 1.44 \times T_{\frac{1}{2}} = \frac{1}{\lambda}$$

Since in radioactive decay not all radioatoms decay at the same time, each individual radioatom exists for a different period of time. Average life is therefore the mean of all these intervals for which individual radioatoms remain in existence.

### Biological Half-Life

Under many circumstances, the disappearance with time of a biochemical or drug in a biological system (*e.g.*, thyroid, liver, lungs, bone, blood, plasma or reticuloendothelial cells) can be described by an exponential law similar to that which holds true for radionuclidic decay. In this case, the disappearance of the biochemical or drug is accomplished through metabolism, excretion, simple diffusion, or some other ill-defined biomechanism. The amount $M_t$ of the biochemical or drug present in the biological system at time t can be determined by an equation similar to that used to determine radioactivity $R_t$ (Equation 4), provided the original amount ($M_0$) of the biochemical or drug and the probability of its disappearance, $\lambda_{Bio}$, in the biological system under consideration is known. In that case,

$$M_t = M_0 e^{-(\lambda_{Bio} \cdot t)} \qquad [7]$$

The disappearance probability, $\lambda_{Bio}$, is related to the biological half-life $T_{\frac{1}{2}}$ (Bio), by an equation similar to equation 5:

$$\lambda_{Bio} \cdot T_{\frac{1}{2}}(Bio) = 0.693 \qquad [8a]$$

The biological half-life is defined as the interval during which a given amount of a biochemical or drug in a biological system is reduced to one-half its original value.

Often the disappearance of a biochemical or drug in the biological system cannot be described by equation 7. In these cases the disappearance of the biochemical or the drug can be described by a sum of several exponential terms as follows:

$$M_t = M_0(A_1 e^{-\lambda 1_{Bio} \cdot t} + A_2 e^{-\lambda 2_{Bio} \cdot t} + \ldots\ldots) \qquad [8b]$$

where $A_1$, $A_2$ . . . and $\lambda 1_{Bio}$, $\lambda 2_{Bio}$ . . . are constants and are determined experimentally for a given drug in a given biological system.

### Effective Half-Life

In nuclear medicine one is interested in the disappearance with time of radioactive substances or drugs from biological systems as a result of both physical decay and biological elimination (metabolism, diffusion or excretion). In this case, therefore, the probability of disappearance of a radiopharmaceutical from a biological system is the sum of both the physical and biological disappearance probabilities, *i.e.:*

$$\lambda + \lambda_{Bio} = \lambda_{eff} \qquad [9]$$

where $\lambda_{eff}$ is the effective disappearance probability for a radioactive substance or drug. If one converts the various disappearance constants in equation 9 to their respective half-lives, equation 9 becomes

$$\frac{0.693}{T_{\frac{1}{2}}} + \frac{0.693}{T_{\frac{1}{2}}(Bio)} = \frac{0.693}{T_{\frac{1}{2}}(eff)}$$

or,

$$\frac{1}{T_{\frac{1}{2}}} + \frac{1}{T_{\frac{1}{2}}(Bio)} = \frac{1}{T_{\frac{1}{2}}(eff)} \qquad [10]$$

From this relationship (Eq. 10), one can determine the effective half-life of a radiopharmaceutical if the physical half-life of the radionuclide and the biological half-life of the drug (or chemical) are known. As a rule, however, $T_{\frac{1}{2}}(eff)$ will always be less than or equal to the shorter of the two, $T_{\frac{1}{2}}$ and $T_{\frac{1}{2}}(Bio)$.

*Examples:*

(1) The biological half-life of iodine in human thyroid is about 64 days, and the physical half-life of $^{131}I$ is 8 days. Determine the effective half-life of $^{131}I$ in the thyroid.

Given:

$$T_{\frac{1}{2}} = 8 \text{ days}, \; T_{\frac{1}{2}}(\text{Bio}) = 64 \text{ days}$$

Therefore, from equation 10,

$$\frac{1}{T_{\frac{1}{2}}(\text{eff})} = \frac{1}{8} + \frac{1}{64} = \frac{9}{64}$$

or

$$T_{\frac{1}{2}}(\text{eff}) = \frac{64}{9} = 7.1 \text{ days}$$

(2) Xenon-133, a radioactive inert gas, is used for lung-function studies. Its physical half-life is 5.3 days, and its biological half-life in the lungs is about 0.35 minute. Determine the effective half-life of $^{133}$Xe in the lungs. Given:

$$T_{\frac{1}{2}} = 5.3 \text{ days} = 5.3 \times 24 \times 60 \text{ min}$$

$$= 7632 \text{ min}$$

$$T_{\frac{1}{2}}(\text{Bio}) = 0.35 \text{ min}$$

Therefore, from equation 10,

$$\frac{1}{T_{\frac{1}{2}}(\text{eff})} = \frac{1}{7632} + \frac{1}{0.35}$$

or,

$$T_{\frac{1}{2}}(\text{eff}) = 0.35 \text{ min}$$

In this example, $T_{\frac{1}{2}}(\text{eff}) = T_{\frac{1}{2}}(\text{Bio})$. Actually, whenever one of the half-lives, $T_{\frac{1}{2}}$ or $T_{\frac{1}{2}}(\text{Bio})$, is very large (10 times or more) in comparison to the other, then $T_{\frac{1}{2}}(\text{eff})$ is equal to the other half-life [$T_{\frac{1}{2}}(\text{Bio})$ or $T_{\frac{1}{2}}$].

In mathematical terms,

$$\text{when } T_{\frac{1}{2}} \gg T_{\frac{1}{2}}(\text{Bio}), \text{ then } T_{\frac{1}{2}}(\text{eff}) \approx T_{\frac{1}{2}}(\text{Bio})$$

or,

$$\text{when } T_{\frac{1}{2}}(\text{Bio}) \gg T_{\frac{1}{2}}, \text{ then } T_{\frac{1}{2}}(\text{eff}) \approx T_{\frac{1}{2}}$$

## STATISTICS OF RADIOACTIVE DECAY

Although $\lambda$ is defined here as the probability of decay per unit time per atom, in earlier discussions it has been tacitly assumed that the number of radioatoms which decay in time t can be accurately determined. This is not so, nor is there any way to predict exactly which particular atom will decay at a given moment. In practice, the number of radioatoms which decay in a given time t fluctuate around an average value, and for this reason the equations described in the preceding sections express only the relationships of the average values of $N_t$ and $R_t$.

Because of these statistical variations in nuclear measurements, it is essential to know what contribution these fluctuations make to the total error of a measurement. Barring blunders, error in any measurement has two causes, systematic and random. Systematic errors are caused by the use of miscalibrated instruments and will always either underestimate a measurement or overestimate it depending on which way the error has been made in calibration. Random errors are caused by some uncontrollable factors. Therefore, when a number of repeat measurements are made, these tend to be randomly distributed. The term, accuracy, is used to describe systematic errors whereas precision is used to describe random errors. In radioactivity measurements, random errors which are the result of the natural decay process dominate over the systematic errors and therefore are taken into account only. I should point out here, though, that this discussion of errors is not limited only to counting but is of importance in imaging as well.

***Poisson Distribution, Standard Deviation, and Percent Standard Deviation:*** The probability that a given number of disintegrations (N) will actually occur in a radioactive sample in time t, when the mean number of disintegrations for that sample is $\overline{N}$, is determined by the Poisson distribution. The mathematical or graphic description of Poisson distribution is complicated. Therefore, it is generally approximated with a Gaussian or Standard distribution which is shown in Fig. 3–3. A Gaussian distribution which is completely determined by two parameters, mean $\overline{N}$ and standard deviation $\sigma$, is given by the following expression

$$P(N) = \frac{1}{\sigma\sqrt{2\pi}} \cdot e^{-\frac{1}{2}\left(\frac{N - \overline{N}}{\sigma}\right)^2}$$

where P(N) is the probability for observing N disintegrations, $\overline{N}$ is the average number of disintegrations and $\sigma$ is the standard deviation. Standard deviation determines the spread or width of a Gaussian (or Poisson) distribution and is commonly employed as a measure of random error, or precision, of a radioactive measurement.

A range of $N - \sigma$ to $N + \sigma$ (1 S.D.) implies that the probability of a given measurement lying within this range is about 68%. When the range is extended to include 2 S.D.s ($N - 2\sigma$ to $N + 2\sigma$), the probability increases to 95%. Further extension of the range to include 3 S.D.s ($N - 3\sigma$ or $N + 3\sigma$) increases the probability to 99%.

An important characteristic of Poisson distribution is the elation of its S.D. to the mean number of disintegrations $\overline{N}$ as follows:

$$\sigma = \sqrt{\overline{N}} \qquad [11]$$

For large values of $\overline{N}$, this equation allows a rapid determination of the standard deviation of a given measurement if we replace $\overline{N}$ by N.

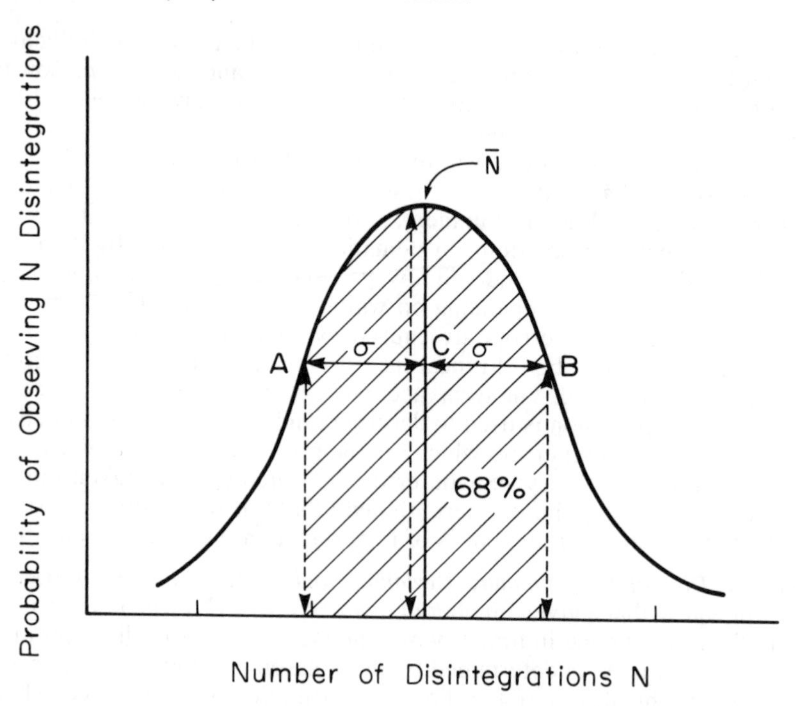


**Fig. 3–3.** Poisson distribution (for large values of N, it is similar to a Gaussian distribution) showing the random nature of radioactivity. The width $\sigma$ (AC or BC) is known as standard deviation (S.D.). A range of N + $\sigma$ to N − $\sigma$ (A to B) covers about 68% of the total area of distribution curve.

A more useful index of statistical error or precision of a measurement than standard deviation is percent standard deviation (% S.D.). The % S.D. of a measurement is determined as follows:

$$\% \text{ S.D.} = \frac{\sigma}{N} \times 100 = \frac{\sqrt{N}}{N} \times 100 = \frac{100}{\sqrt{N}} \qquad [12]$$

This equation states that the precision of a given radioactive measurement can be increased by increasing the observed number of disintegrations. In practical terms this means that if we measure a radioactive sample for one minute and obtain 100 disintegrations during this time, then its % S.D., as determined from the above relationship, will be equal to $\frac{100}{\sqrt{100}}$ or 10%. However, if we measure the sample long enough (approximately 100 times longer) so that there are 10,000 disintegrations during this interval, the % S.D. in this case will be equal to $\frac{100}{\sqrt{10,000}}$ or 1%. In other words, by increasing the time of counting (approximately 100 times) we have decreased the percent error in our results (though by a factor of 10 only). When the time of counting is limited, a compromise has to be made between the length of counting time and precision.

***Propagation of Statistical Errors:*** In many nuclear applications we have to add, subtract, multiply or divide two or more measurements. How does one determine the precision of the final result? To answer this question let us assume that N1 and N2 are two measurements from which the final result N is obtained by either adding, subtracting, dividing or multiplying N1 and N2. Without going into detail, then, the % S.D. of N is given by the following expressions:

$$\text{When adding: } \% \text{ S.D.} = \frac{100}{\sqrt{N1 + N2}} \quad [13]$$

$$\text{When subtracting: } \% \text{ S.D.} = \frac{100 \cdot \sqrt{N1 + N2}}{N1 - N2} \quad [14]$$

$$\text{When multiplying or dividing: } \% \text{ S.D.} = \sqrt{(\% \text{ S.D. of N1})^2 + (\% \text{ S.D. of N2})^2} \quad [15]$$

*Error in count rate:* Count rate R of a sample is determined by dividing the number of counts N, observed in a time interval t by the time interval t or

$$R = \frac{N}{t} \text{ or } N = R \cdot t \quad (16)$$

Since the error in the measurement of time t is generally very, very small, the error in R mainly results from the error in N. Therefore

$$\sigma_R = \frac{\sigma_N}{t} = \frac{\sqrt{N}}{t} \quad (17)$$

substituting $N = R \cdot t$ from equation 16, we get

$$\sigma_R = \frac{\sqrt{R \cdot t}}{t} = \sqrt{\frac{R}{t}} \quad (18)$$

To reduce the error of a count rate R, one has to increase the averaging time t. The $\%\sigma_R$ of a count rate is given by the following expression:

$$\%\sigma_R = \frac{\sigma_R \times 100}{R} = \sqrt{\frac{R}{t}} \cdot \frac{100}{R} = \frac{100}{\sqrt{R \cdot t}} = \frac{100}{\sqrt{N}} \quad (19)$$

***Room Background:*** Even when there is no radioactive sample near a radiation detector, it will still record a certain number of disintegrations known as room background. This is due to the presence of minute amounts of radioactivity in the surrounding earth and/or building material. Part of the room background is also contributed by the high-energy radiations of cosmic rays. Room background can be reduced to a low level by shielding the radiation detector. However, it cannot be reduced to zero because of the presence of trace amounts of radioactive substances in the detector and shielding materials themselves, as well as because of the inability to completely shield against high-energy background radiation. In situations where the sample count rate is very low (less than 10 times of background) it is important to take into consideration the room background as well as its effect on the precision of the final result.

*Examples:*

(1) In the thyroid uptake measurement of a patient, it is found that the neck activity gives 900 counts per minute, whereas the standard is 2500 counts per minute. Calculate the % uptake by thyroid and its precision.

$$\% \text{ uptake} = \frac{\text{counts in neck}}{\text{counts in standard}} \times 100$$

$$= \frac{900}{2500} \times 100 = 36\%$$

Using equation 12,

$$\% \text{ S.D. in neck counts} = \frac{100}{\sqrt{900}} = 3.3\%$$

and,

$$\% \text{ S.D. in standard counts} = \frac{100}{\sqrt{2500}} = 2\%$$

Using equation 15,

$$\% \text{ S.D. of thyroid uptake} = \sqrt{3.3^2 + 2^2}$$

$$= \sqrt{11 + 4} = 4\%$$

Therefore,

$$\text{Thyroid uptake} = (36 \pm 4\%)\%$$

Since 4% of 36 is 1.4,

$$\text{Thyroid uptake} = (36 \pm 1.4)\%$$

(2) A radioactive sample registers a total (sample + background) of 3200 counts in 1 minute. The room background is 1000 counts per minute. Calculate the net counts of the sample and the % S.D. of the total counts, background and the net counts, respectively.

$$\text{Net counts in the sample} = \text{total counts} - \text{background counts} = 3200 - 1000 = 2200 \text{ counts}$$

Using equation 12,

$$\% \text{ S.D. of the total counts} = \frac{100}{\sqrt{3200}} = 1.7\%$$

and,

$$\% \text{ S.D. of the background counts} = \frac{100}{\sqrt{1000}} = 3.1\%$$

Using equation 14,

$$\% \text{ S.D. of the net counts} = \frac{100\sqrt{3200 + 1000}}{3200 - 1000} = \frac{100\sqrt{4200}}{2200} = 2.9\%$$

This clearly demonstrates the manner in which the room background degrades the precision of a radioactive measurement.

## PROBLEMS

1. Determine the radioactivity of a sample disintegrating at a rate of 10,000 disintegrations per second in (a) millicuries and (b) megabecquerels.
2. What is a 10 mCi dose of $^{99m}$Tc radiopharmaceutical equivalent to in SI units?
3. If the decay constant, $\lambda$, of the radionuclide in problem 1 is (a) 0.1 per second and (b) 0.1 per hour, how many radioatoms are present in this sample?
4. Determine the radioactivity of a $^{99m}$Tc sample (370 MBq at 0 time) after (a) 9 hours, (b) 12 hours, and (c) 60 hours of decay.
5. How long will it take for the radioactivity of a sample to decay to its 25% value if the radionuclide is (a) $^{201}$Tl, (b) $^{99}$Mo, or (c) $^{67}$Ga?
6. The biological half-life of $^{99m}$Tc macroaggregated albumin in the lungs is 6 hours. What is the effective half-life of this agent in the lungs?
7. How many counts are needed to make the standard deviation equal to 1%?
8. A series of 100 one-minute measurements were made, giving an average count-rate of 1600 cpm. Of these, how many are in the range of 1600 ± 80?
9. A 99% confidence level of counts taken at the rate of 1200 cpm for 3 minutes is ______ counts.
10. If one obtains 10,000 counts in 5 minutes, what is the standard deviation of the count rate?

# 4

# Production of Radionuclides

In 1896, Henry Becquerel discovered that uranium was radioactive. Soon after, other naturally occurring radionuclides such as radium and polonium were discovered. Most of the naturally occurring radionuclides have long half-lives (greater than 1,000 years) and are not used in nuclear medicine. The radionuclides most commonly used in nuclear medicine are man-made, produced by three basic methods:

1. Irradiation of stable nuclides in a reactor (reactor-produced).
2. Irradiation of stable nuclides in an accelerator or cyclotron (accelerator- or cyclotron-produced).
3. Fission of heavier nuclides (fission-produced).

## METHODS OF RADIONUCLIDE PRODUCTION

### Reactor-Produced Radionuclides

A nuclear reactor (for a description of a nuclear reactor, see Fission-Produced Radionuclides, p. 49) is a source of a large number of thermal neutrons. Thermal neutrons are those neutrons whose kinetic energy is very small (~.025 eV, which is the kinetic energy of atoms or molecules at room temperature). At these energies, neutrons can be easily captured by stable nuclides since neutrons which are neutral particles do not experience the repulsive coulomb forces of the positively charged nucleus. The capture reaction of neutrons with a given nuclide ${}^{A}_{Z}X$ is represented by either of the following notations:

(i) $${}^{A}_{Z}X + {}^{1}_{0}n \rightarrow {}^{A+1}_{Z}X + \gamma\text{-rays}$$

(ii) $${}^{A}_{Z}X\,(n, \gamma)\,{}^{A+1}_{Z}X$$

In (i), the reactants are to the left of the arrow and the products of the reaction are to the right. The first equation is written as a chemical reaction;

the second equation is a short notation for the same. It can be seen that in the above nuclear reaction the atomic number (and, therefore, the chemical nature or element) of the resulting nuclide ($^{A+1}_{Z}X$) does not change; only the mass number, A, increases by 1. Since in this reaction a neutron is being added, the resulting nuclide (if radioactive) quite often decays through $\beta$-emission. I say "if radioactive" because for many neutron capture reactions the resulting nuclide is stable, *e.g.*, $^{12}_{6}C$ (n, $\gamma$) $^{13}_{6}C$. In this example, the $^{13}_{6}C$ nuclide is a stable nuclide. Another feature of reactor-produced nuclides is that these, in general, are not carrier-free. In a carrier-free sample only the desired radionuclide is present without contamination from its other isotopes. A sample of $^{131}_{53}I$ can be called carrier-free only if no other stable isotope or radioisotope of iodine is present in the sample.

Some of the reactor-produced radionuclides, and the nuclear reactions producing them, used in nuclear medicine are given below:

(i) $$^{50}_{24}Cr + ^{1}_{0}n \rightarrow ^{51}_{24}Cr + \gamma$$

$^{51}Cr$ is used for labeling red blood cells and spleen scanning.

(ii) $$^{98}_{42}Mo + ^{1}_{0}n \rightarrow ^{99}_{42}Mo + \gamma$$

$^{99}Mo$ is the source of $^{99m}Tc$, which is so commonly used in nuclear medicine.

(iii) $$^{132}_{54}Xe + ^{1}_{0}n \rightarrow ^{133}_{54}Xe + \gamma$$

$^{133}Xe$ is used for lung ventilation studies.

### Accelerator- or Cyclotron-Produced Radionuclides

An accelerator or cyclotron is the source of a large number of high-energy (MeV range) charged particles such as p (protons), $^{2}_{1}D$ (deutrons), $^{3}_{2}He$ (helium 3) and $^{4}_{2}He$ (alpha) particles. The classification of an accelerator or a cyclotron depends on the way in which these charged particles are accelerated and is not relevant here. The probability of a nuclear reaction occurring with charged particles is highly dependent on the energy of the bombarding particles. For each charged particle and target there is a threshold energy below which the nuclear reaction does not occur at all. This is due to the repulsive coulomb forces between the positively charged particle and the positively charged target nuclide. Generally, the threshold energy is in the MeV range. The most common reactions for protons are:

(i) $$^{A}_{Z}X + ^{1}_{1}p \rightarrow {}_{Z+1}^{A}Y + n$$

or

$$^{A}_{Z}X \text{ (p, n) } {}_{Z+1}^{A}Y$$

(ii) $$^{A}_{Z}X + ^{1}_{1}p \rightarrow ^{A-1}_{Z+1}Y + 2n$$

or

$$^{A}_{Z}X \text{ (p, 2n) } ^{A-1}_{Z+1}Y$$

The most common reactions for deutrons, ${}^{2}_{1}D$ (also known as heavy hydrogen), are (in short notations):

(iii) $${}^{A}_{Z}X\ ({}^{2}_{1}D,\ n)\ {}^{A+1}_{Z+1}Y$$

(iv) $${}^{A}_{Z}X\ ({}^{2}_{1}D,\ p)\ {}^{A+1}_{Z}X$$

(v) $${}^{A}_{Z}X\ ({}^{2}_{1}D,\ 2n)\ {}^{A}_{Z+1}Y$$

Common nuclear reactions for ${}^{3}_{2}He$ particles are:

(vi) $${}^{A}_{Z}X\ ({}^{3}_{2}He,\ n)\ {}^{A+2}_{Z+2}Y$$

(vii) $${}^{A}_{Z}X\ ({}^{3}_{2}He,\ p)\ {}^{A+2}_{Z+1}Y$$

Common reactions for $\alpha$ (${}^{4}_{2}He$) particles are:

(viii) $${}^{A}_{Z}X\ ({}^{4}_{2}He,\ n)\ {}^{A+3}_{Z+2}Y$$

(ix) $${}^{A}_{Z}X\ ({}^{4}_{2}He,\ 2n)\ {}^{A+2}_{Z+2}Y$$

Most of the above reactions occur in the range of 5 to 30 MeV. As the energy of the bombarding particles further increases, other nuclear reactions occur. Sometimes these additional reactions may also be useful for producing radionuclides. Some of the radionuclides used routinely in nuclear medicine and produced in an accelerator or cyclotron are given below:

(i) $${}^{16}_{8}O + {}^{3}_{2}He \rightarrow {}^{18}_{9}F + p$$

${}^{18}F$ is used for labeling radiopharmaceuticals for PET imaging.

(ii) $${}^{68}_{30}Zn + p \rightarrow {}^{67}_{31}Ga + 2n$$

${}^{67}Ga$ is widely used for soft tumor and occult abscess detection.

In the above examples, the radionuclide of interest is formed directly as a result of a particular nuclear reaction. Sometimes, radionuclide of interest may be formed indirectly, by the decay of another radionuclide which is formed first with a nuclear reaction. Two examples of these indirect methods are the production of radionuclides ${}^{123}I$ and ${}^{201}Tl$:

(i) Production of ${}^{123}I$

(ii) Nuclear Reaction $\quad {}^{122}_{52}Te + {}^{4}_{2}He \rightarrow {}^{123}_{54}Xe + 3n$ (2 hr, half-life)

(iii) Decay $\quad {}^{123}_{54}Xe \xrightarrow{E.C.,\ \beta^{+}} {}^{123}_{53}I$ (13 hr, half-life)

(2) Production of ${}^{201}Tl$

(i) Nuclear Reaction, $\quad {}^{203}_{81}Tl + p \rightarrow {}^{201}_{82}Pb + 3n$ (9.4 hr, half-life)

(ii) Decay, $\quad {}^{201}_{82}Pb \xrightarrow{E.c.} {}^{201}_{81}Tl$ (73 hr, half-life)

Since, in charged-particle nuclear reactions, the resultant radionuclide generally has an atomic number different from that of the target nuclide, one can chemically separate the two. Therefore, the radionuclides produced by charged particle reactions are generally carrier-free. Also since in these reactions protons are added to a nuclide, these are generally $\beta^+$- or K-capturing radionuclides.

### Fission-Produced Radionuclides

Soon after the discovery of radioactivity, naturally occurring radioactive nuclides such as $^{226}_{88}$Ra, $^{232}_{90}$Th, or $^{210}_{84}$Po were found to be good sources of $\alpha$ particles. The reactions of these $\alpha$ particles produced neutrons by $^{A}_{Z}$X ($\alpha$,n), $^{A+3}_{Z+2}$Y. When the reactions of the neutrons thus generated were systematically studied, a surprising discovery was made. For many heavier nuclei (A ~ 200) it was found that capture of a neutron, instead of producing a heavier radionuclide, resulted in the production of several radionuclides whose mass numbers were about one-half that of the target nuclide. For example, in the case of $^{235}$U,

$$^{235}_{92}\mathrm{U} + ^{1}_{0}\mathrm{n} \rightarrow ^{236}_{92}\mathrm{U} + \gamma \text{ seldom occurs.}$$

Instead, $^{235}_{92}\mathrm{U} + ^{1}_{0}\mathrm{n} \rightarrow ^{141}_{56}\mathrm{Ba} + ^{91}_{36}\mathrm{Kr} + 4\,^{1}_{0}\mathrm{n}$ is a much more frequent reaction.

This process of splitting a heavier nucleus into two small nuclei is called fission. Barium and krypton are not the only elements formed in fission. Actually, every element from zinc (Z = 30) to dysprosium (Z = 66) has been identified in fission reaction. Besides the production of radionuclides of intermediate elements (Z = 30 to 66) during fission, another important result is the production of a large number of neutrons (in the above example, 4). A neutron initiates the fission by being captured, yet more than one neutron is produced during fission. These extra neutrons can then cause further fission, thereby producing an even larger number of neutrons. This process is a chain reaction and will theoretically continue until the supply of fissionable material is exhausted. An uncontrolled chain reaction of fissionable material is called an atomic bomb. A controlled chain reaction, however, which is a very good source of a large number of neutrons and of energy (for producing electricity), is known as a nuclear reactor. $^{131}$I, so commonly used in nuclear medicine, is produced by fission. Another example of a fission-produced radionuclide is $^{99}$Mo, the parent radionuclide of $^{99m}$Tc. Like cyclotron- or accelerator-produced radionuclides, fission-produced radionuclides are also generally carrier-free.

## GENERAL CONSIDERATIONS IN THE PRODUCTION OF RADIONUCLIDES

The amount of radioactivity $R_t$ produced in time t in a nuclear reaction depends on the following factors:

1. The number of bombarding particles/sec/cm$^2$ known as Flux, I.
2. The total number of the target nuclei irradiated, n × V, when n is the number of the target nuclides in 1 cm$^3$ and V is the volume of the target material being irradiated.
3. The time of irradiation, t.
4. The half-life or the decay constant of the radionuclide produced, *i.e.*, $T\frac{1}{2}$ or $\lambda$.
5. The probability of the given nuclear reaction, called the cross-section $\sigma$. The unit for cross section is a barn, which is equal to $10^{-24}$ cm$^2$.

In the case of neutron capture reactions (reactor-produced radionuclides) the formula relating the above factors to the activity produced after time t is

$$R_t = \sigma \cdot I \cdot n \cdot V \cdot (1 - e^{-\lambda t}) \quad [1]$$

Because of the factor $(1 - e^{-\lambda t})$, it does not pay to irradiate a target for more than one half-life of the desired radionuclide. In cases where the half-life of the desired radionuclide is sufficiently long (in days), the above equation can be reduced to a simpler form:

$$R_t = \sigma \cdot I \cdot n \cdot V \cdot \lambda \cdot t \quad [2]$$

A similar equation can be written for charged-particle nuclear reactions. In practice, however, the flux of charged particles is measured as $\mu$amp (a unit of electric current) instead of the number per sec/per cm$^2$. In this case, too, the amount of activity produced will depend on the five factors given above. In general, the values quoted for the activity produced by charged-particle reactions are in units of mCi/$\mu$amp/hr and are known as the yield of a particular reaction. The higher the yield for a particular nuclear reaction, the easier it is to produce large quantities of the given radionuclide. In selecting the best method of radionuclide production, one has to take into account (1) yield, which is an economical consideration; and (2) purity and the specific activity of the radionuclide, which are biological or scientific considerations and, therefore, depend on the particular use of the radionuclide.

## PRODUCTION OF SHORT-LIVED RADIONUCLIDES, USING A GENERATOR

Because of the reduction in radiation dose to the patient, short-lived radionuclides are often the agents of choice in nuclear medicine. In general, however, the use of short-lived radionuclides entails many problems due to their fast decay. For example, the short half-life of a radionuclide limits the available time for such purposes as processing, transportation, storage and quality control. For this reason $^{18}F$, with a half-life of 100 minutes, does not have widespread use as a bone-scanning agent. Similar limitations apply to the widespread use of other short-lived radionuclides such as $^{11}C$ (20.3 min), $^{13}N$ (10 min) and $^{15}O$ (2 min), which are quite attractive from other considerations. A radionuclide generator described below (also referred to as a "cow") solves some of the above problems and allows the use of short-lived radionuclides at long distances from the site of production, *e.g.*, a cyclotron or reactor.

## Principles of a Generator

A radionuclidic generator is a two- or three-step radioactive series in which a long-lived radionuclide (also called "parent") decays into a short-lived radionuclide (also known as "daughter") of interest.

The following are some examples of radionuclide generators. The first generator system is the most common in use today. The second generator system is useful for geographically remote places where frequent deliveries of the radionuclides are a problem. The third generator is under development and is receiving attention because of its special advantages for positron tomography. The fourth generator produces $^{81m}Kr$, which is used in lung ventilation studies sometimes.

| | | | | |
|---|---|---|---|---|
| (i) | $^{99}Mo \rightarrow$ | $^{99m}Tc \rightarrow$ | $^{99}Tc \rightarrow$ | $^{99}Ru$ |
| Half-life: | 67 hr | 6 hr | long | stable |
| (ii) | $^{113}Sn \rightarrow$ | $^{113m}In \rightarrow$ | $^{113}In$ | |
| Half-life: | 115 d | 1.67 hr | stable | |
| (iii) | $^{68}Ge \rightarrow$ | $^{68}Ga \rightarrow$ | $^{68}Zn$ | |
| Half-life: | 275 d | 1.1 hr | stable | |
| (iv) | $^{81}Rb \rightarrow$ | $^{81m}Kr \rightarrow$ | $^{81}Kr$ | |
| Half-life: | 4.7 hr | 13 s | stable | |

In a radioactive series, the daughter radionuclide is being continuously produced by the decay of the parent radionuclide as well as being continuously destroyed by its own decay. If the half-life of the parent radionuclide is longer than the half-life of the daughter radionuclide, an important phenomenon occurs which is the basis of the generators presently used in nuclear medicine. Under this condition (*i.e.*, $T_{\frac{1}{2}}$ parent > $T_{\frac{1}{2}}$ daughter), and in due course an equilibrium is established between the parent and daughter radionuclides. In the state of equilibrium the ratio of the amounts (number of radionuclei present) of the two radionuclides becomes constant. The two radioactivities also maintain a constant ratio (this ratio is in general very close to unity) with time even though the half-lives of the two radionuclides are quite different. In effect, the daughter radioactivity decays with an apparent half-life of the parent radionuclide rather than its own. For example, in a $^{99}Mo$-$^{99m}Tc$ generator, the $^{99m}Tc$ radioactivity in equilibrium with $^{99}Mo$ decays with a half-life of 67 hr rather than 6 hr. The growth and decay of the daughter activity in a generator can be exactly predicted using the decay laws given in Chapter 3. Rather than going through complicated mathematics, we have shown this relationship graphically in Figs. 4–1 and 4–2

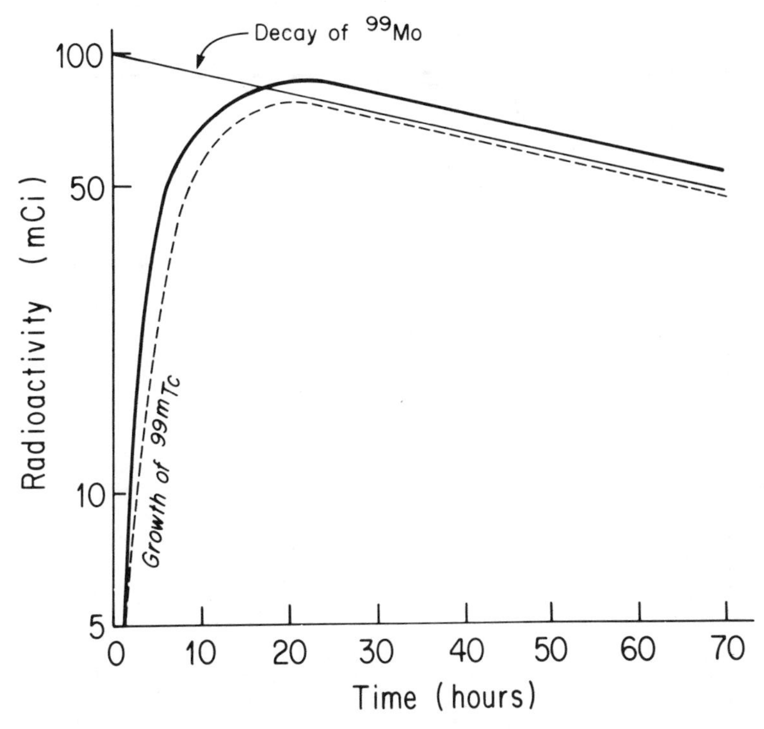


**Fig. 4–1.** Growth and decay of $^{99m}$Tc radioactivity in the decay of $^{99}$Mo. It takes about 24 hours for equilibrium to establish. For the solid curve, it is assumed that all disintegrations of $^{99}$Mo produce metastable technetium, $^{99m}$Tc. The broken curve represents the actual radioactivity of $^{99m}$Tc since only 92% disintegrations of $^{99}$Mo produce $^{99m}$Tc. The remaining 8% of $^{99}$Mo disintegrations produce the ground state, $^{99}$Tc, via short lived isomers.

for the two most common generators. In both cases it can be seen that it takes about 4 daughter half-lives to reach equilibrium. For a $^{99}$Mo-$^{99m}$Tc generator, it is about 24 hr, whereas for a $^{113}$Sn-$^{113m}$In generator it is about 6.5 hr. It is also evident from Figs. 4–1 and 4–2 that the growth of the daughter radioactivity is not linear with time. Instead, it takes about 1 daughter half-life to reach 50%, 2 daughter half-lives to reach 75% and 3 daughter half-lives to reach 87% of the equilibrium level.

Once equilibrium has been achieved between the parent and daughter radioactivities, it can be disturbed only by chemical separation (*i.e.*, "milking") of the two radionuclides. After chemical separation, the daughter radioactivity again grows and re-establishes an equilibrium (in about 4 daughter half-lives) with the parent radioactivity, although at a new level. In other words, one has a fresh supply of the daughter radionuclide available for milking each 4 daughter half-lives after the previous milking. However, one does not necessarily have to wait 4 daughter half-lives before milking. Four daughter half-life intervals provide the maximum obtainable yield from a generator. If an emergency supply of the daughter radionuclide is needed, the generator can be remilked sooner. For example, after 1 daughter half-

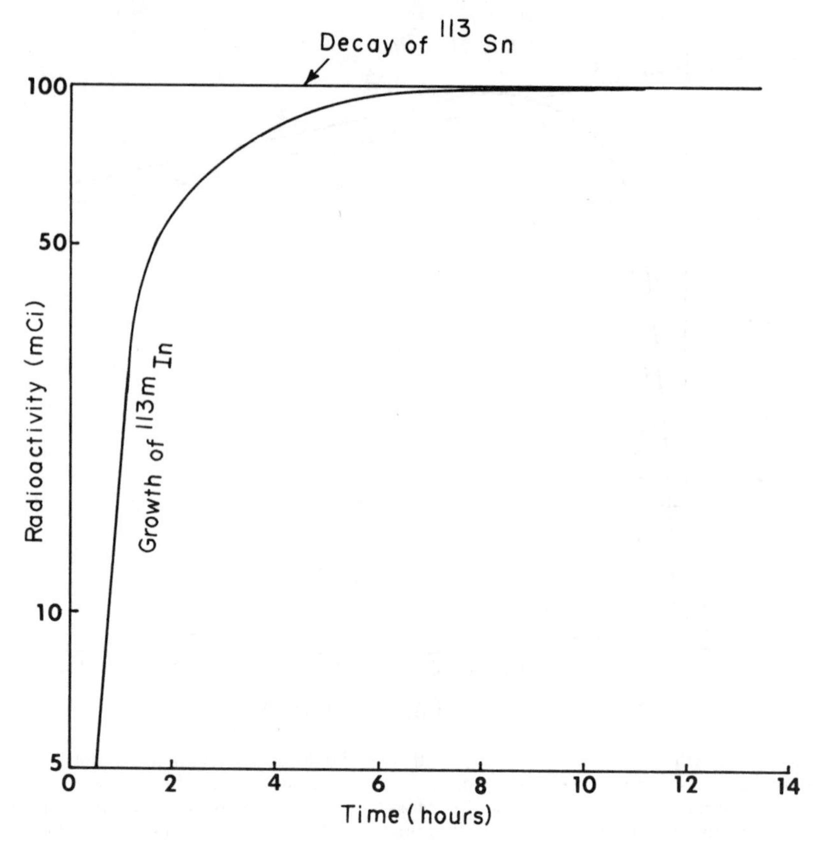


**Fig. 4–2.** Growth and decay of $^{113m}$In radioactivity in the decay of $^{113}$Sn. It takes approximately 6.5 hr for equilibrium to be established.

life, remilking will provide about 50% of the maximum obtainable radioactivity of the daughter radionuclide.

The state of equilibrium between the parent and daughter radionuclides is sometimes classified in two categories, transient and secular. When the half-life of a parent radionuclide is not long in comparison with the half-life of the daughter radionuclide, the equilibrium is called transient. $^{99}$Mo-$^{99m}$Tc and $^{87}$Y-$^{87m}$Sr generators are two examples of transient equilibrium. On the other hand, when the half-life of the parent radionuclide is much longer than that of the daughter radionuclide, the equilibrium is termed secular. Two examples of secular equilibrium are $^{113}$Sn-$^{113m}$In and $^{226}$Ra-$^{222}$Rn generators. In secular equilibrium the activities of the parent and daughter radionuclides become nearly equal, whereas in transient equilibrium, the daughter radioactivity is slightly higher than the parent radioactivity. This distinction between secular and transient equilibrium is of academic interest only, as it carries no practical significance.

Since in the case of a generator a daughter radionuclide is chemically separated from the parent, the daughter radionuclide is produced almost carrier-free.

### Description of a Typical Generator

A typical generator is comprised of a glass column filled with a suitable exchange material such as alumina ($Al_2O_3$). The bottom of the glass column is fitted with a porous glass disk so as to retain the alumina in the column (Fig. 4–3). The parent radionuclide in equilibrium with its daughter is firmly absorbed on the top of the alumina. The daughter radionuclide is separated (eluted or milked) from its parent radionuclide by passing a special liquid (eluting solution) through the column at a suitable flow rate. The daughter is dissolved in the eluting solution, whereas the parent is retained by the column.

In a typical $^{99}$Mo-$^{99m}$Tc generator, the column is filled with alumina, $^{99}$Mo is part of the molecule, sodium molybdate, and the eluting solution is oxident free physiologic saline (0.9% sodium chloride solution). The technetium radioactivity elutes in the form of sodium pertechnetate (Na $^{99m}$Tc $O_4$). Molybdenum-99 used in these generators is produced either by irradiation of $^{98}$Mo with neutrons or by fission of $^{235}$U in a reactor. Molybdenum-99 produced in fission reactions is essentially carrier-free and therefore has very high specific activity. On the other hand, $^{99}$Mo produced by neutron irradiation of $^{98}$Mo is generally of low specific activity. The other difference

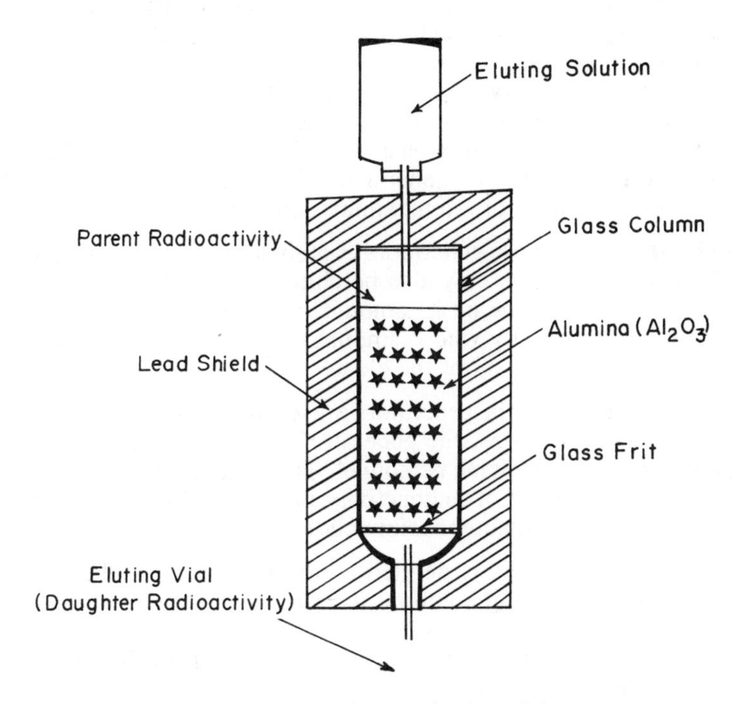


**Fig. 4–3.** Outline of a typical radionuclide generator.

between these two types of $^{99}$Mo is the presence of very minute amounts of other radionuclides (radionuclidic impurities). In the $^{99}$Mo produced by neutron irradiation, the most common radionuclidic impurities are $^{134}$Cs, $^{60}$Co, $^{86}$Rb, $^{124}$Sb and $^{95}$Zr, whereas in the fission produced $^{99}$Mo, the most common radionuclidic impurities are $^{131}$I, $^{132}$I, $^{89}$Sr, $^{90}$Sr and $^{103}$Ru. Since some of these radionuclides may be eluted with $^{99m}$Tc during the milking of the generator, their amounts in the $^{99}$Mo sample should be as small as possible.

For any generator system to be of practical use in nuclear medicine, the milking process should be convenient and rapid. In the selection of a suitable generator one has to consider several other factors such as efficiency, amount of parent breakthrough, radiation shielding and specific concentration. These factors are discussed below with special reference to the $^{99}$Mo-$^{99m}$Tc generator system.

***Efficiency:*** The efficiency of a generator is defined in the following manner:

$$\text{Efficiency} = \frac{\text{Amount of activity eluted}}{\text{Total daughter activity on the column}} \times 100$$

This is also referred to as the yield of the generator. Present-day $^{99}$Mo-$^{99m}$Tc generators give a high yield of $^{99m}$Tc, generally between 70 and 90%.

***Parent Breakthrough:*** This is the amount of the parent radioactivity which is eluted with the daughter radioactivity. The amount should be as small as possible because any contamination by a long-lived radionuclide (in this case the parent) increases the radiation dose without providing any benefit to the patient. The maximum allowable level of breakthrough depends primarily on the radiation dose delivered by this contaminant. For the $^{99}$Mo-$^{99m}$Tc generator, the maximum allowable level is 0.15 $\mu$Ci (5.6 KBq) of $^{99}$Mo for each mCi (37MBq) of $^{99m}$Tc. In general the $^{99}$Mo breakthrough in $^{99}$Mo-$^{99m}$Tc generators is about $\frac{1}{10}$ of the maximum allowable level.

***Chemical Purity:*** Since these generators contain large amounts of alumina, occasionally some aluminium may also be eluted with technetium pertechnetate. Aluminium, depending on its amount, may form colloid. This is an undesirable chemical impurity whose amount should be tested. Its amount should not exceed 10 $\mu$g per ml of the eluate.

***Radiation Shielding:*** Since the generators presently in use routinely contain curie amounts of radionuclides, the generator should be properly shielded for the safety of nuclear-medicine personnel. The exact amount of shielding depends on the total amount of radioactivity present and the energy of the $\gamma$-rays of the parent and daughter radionuclides.

***Specific Concentration:*** This is defined as the number of millicuries per milliliter of the milked eluting solution. This number should be generally high for two reasons:

(1) In dynamic studies, such as cardiac or brain blood flow, it is important to have a small concentrated bolus.

(2) When labeling a variety of pharmaceuticals with the daughter radio-

nuclide, various dilution steps are required. Therefore, to obtain a reasonably high specific concentration of the labeled compound, it is necessary to begin with a very high specific concentration of the radionuclide. Too high a specific concentration of the radioactivity is, however, of no practical value.

Present generators yield up to 100 mCi/ml specific concentration for $^{99m}$Tc.

## PROBLEMS

1. Why do reactor-produced radionuclides generally decay with $\beta$- and accelerator-produced radionuclides with $\beta^+$ and/or electron capture?
2. Why are carrier-free radionuclides desirable in nuclear medicine?
3. Why is the $^{99}$Mo-$^{99m}$Tc radionuclidic generator so popular in nuclear medicine?
4. Why is the parent breakthrough in a generator undesirable and to be kept to a minimum?
5. A radionuclidic generator consisting of $^{132}$Te (half-life = 78 hours) and $^{132}$I (half-life = 3.2 hours) is in transient equilibrium. If the radioactivity of $^{132}$Te at this time is 16 GBq, how much radioactivity of $^{132}$I will be available 156 hours later if no milking took place during this interval?
6. The above generator is completely milked at 156 hours. The $^{132}$I-radioactivity thus obtained is left to decay for 16 hours. How much radioactivity of $^{132}$I remained at this time?

# 5

# Radiopharmaceuticals

In nuclear medicine radionuclides are rarely used in their simplest chemical form. Instead, they are incorporated in a variety of chemical compounds which may be of interest because of their favorable biochemical, physiologic or metabolic properties. A chemical compound tagged with a radionuclide and prepared in a form suitable for human use is known as a radiopharmaceutical. A radiopharmaceutical is used (with a few exceptions discussed at the end of this chapter) to obtain diagnostic information rather than to produce therapeutic results. It is usually administered in tracer quantities in a single dose and produces no pharmacologic effects.

## DESIGN AND DEVELOPMENT OF A RADIOPHARMACEUTICAL

Since a radiopharmaceutical consists of a radionuclide and a biochemical, two considerations apply in designing or developing a radiopharmaceutical, one relating to the radionuclide and the other relating to the biochemical.

The choice of a radionuclide for imaging purposes is chiefly dictated by (1) the necessity of minimizing the radiation dose to the patient and (2) the detection characteristics of present-day nuclear-medicine instrumentation. To minimize the radiation dose to the patient, a radionuclide should have as short a half-life as is compatible with the biological phenomena under study. For example, a radionuclide with a 1-hr half-life, in spite of its smaller radiation dose, cannot be used in studies of physiologic or metabolic functions which span months. A suggested rule of thumb in this connection is that the physical half-life of the radionuclide should be about $0.693 \times T_{obs}$ where $T_{obs}$ is the time interval between the time of administration of the radionuclide and the time at which measurement or scanning is to be performed.

A radionuclide should preferably emit a monochromatic (single-energy) $\gamma$-ray with energy between 100 and 300 keV. The upper limit of the desired energy of the gamma ray is the consequence of the detection characteristics of present day nuclear-medicine instrumentation. As the energy of gamma rays increases, they become more and more penetrating and, therefore, a smaller number of them interact within the detector. This reduces the sensitivity of the system. The concept of sensitivity and its importance in nuclear medicine will be discussed in later chapters. The lower limit of the desired energy of the gamma ray is arrived at from the consideration of attenuation of gamma rays in the patient. Since gamma rays should be able to penetrate the patient's body effectively, their energy has to be high enough to be transmitted out of the patient's body; hence, the lower limit. In addition, a radionuclide should not emit any corpuscular radiation (*e.g.*, $\beta$ particles, conversion electrons) and should be available easily, economically and in an uncontaminated form. Technetium 99m, with its 6-hr half-life and 140-keV $\gamma$-ray emission, comes very close to fulfilling the above requirements. This accounts for its wide use in nuclear medicine.

Besides being nontoxic in the desired amounts the choice of the biochemical or pharmaceutical substance in a radiopharmaceutical is dictated by the requirement that it be distributed or localized in the desired organ or compartment and that the uptake by that organ (or part of the organ) in a normal condition differs substantially from uptake in a pathological condition.

To help in the selection of a suitable biochemical, a wealth of information has been acquired in the field of pharmacology. A number of physiochemical variants determine or affect the distribution and localization of drugs in tissues. Three important determinants in this regard are route of administration, blood flow to the organ or tissues, and extraction by the tissues. Radiopharmaceuticals, with few exceptions, are nearly always administered intravenously, primarily because this is the fastest way to introduce a drug into the circulatory system of the body. Blood flow (which can be severely affected in diseases) essentially determines the fraction of the administered dose that will be delivered to a particular organ or tissue during the first transit. Since blood serves as a carrier for the drug, another property, binding to plasma proteins, plays an important role in the localization of a drug or a chemical in a given tissue. In general, drugs or chemicals strongly bound to plasma proteins localize to a lesser extent in tissues than those which are not so tightly bound to plasma proteins. Extraction of a drug or chemical from circulation and localization in tissue may occur in four ways: simple diffusion, filtration through small pores in the membranes, active transport and phagocytosis. It is evident from Table 5–1 that all of these mechanisms have been utilized in the development of the radiopharmaceuticals.

**Table 5–1. Mechanism of Localization of Radiopharmaceuticals**

| *Mechanism* | *Example* |
|---|---|
| Active Transport | Thyroid uptake and scanning with iodine. |
| Compartmental Localization | Blood pool scanning with human serum albumin, plasma or red blood cell volume determinations. |
| Simple Exchange or Diffusion | Bone scanning with $^{99m}$Tc labeled phosphate compound. |
| Phagocytosis | Liver, spleen and bone-marrow scanning with radiocolloids. |
| Capillary blockade | Lung scanning with macroaggregate (size 8 to 75 $\mu$), organ perfusion studies with intra-arterial injection of macroaggregates. |
| Cell Sequestration | Spleen scanning with damaged red blood cells. |

Once appropriate radionuclides and chemicals have been selected, the following steps are involved in the eventual development of a radiopharmaceutical:

***Chemical Studies:*** These are aimed at establishing the best method of radiolabeling the chemical, define the optimum condition of labeling and in-vitro stability, and determine the nature and the extent of the radiochemical impurities.

***Animal Distribution and Toxicity Studies:*** The main purpose of these studies is to determine biodistribution of labeled material and establish safe amounts (radiation as well as chemical) of the radiochemicals which can be administered to humans without subjecting them to undue risk. Biodistribution establishes the pattern of distribution (major organ or tissues of uptake) of the radioactivity at different times after the administration of the radiochemical in animals considered normal (control) and those in which the appropriate pathologic condition has been induced. From these, one estimates the optimum time for imaging after administration of the radiopharmaceutical.

***Human or Clinical Studies:*** Since biodistribution of a radiopharmaceutical in animals may be different from that in humans, initial studies (Phase I) in only a small number of humans are performed to establish the distribution patterns, clearance time, mode of excretion and optimum imaging time for the radiopharmaceutical. In Phase II, these studies are extended to include patients with known diseases and provide further evidence of safety and initial proof of the diagnostic or therapeutic efficacy. Finally (Phase III), a large series of patients are studied which establishes the overall usefulness, *i.e.*, safety and efficacy, of the agent.

## QUALITY CONTROL

Since all radiopharmaceuticals are intended eventually for human use, strict quality control is very important. To assure optimum quality, the following properties of a radiopharmaceutical must be considered:

***Radionuclidic Purity:*** Ideally the radiopharmaceutical should contain only the desired radionuclide. Often, however, it is not possible to avoid some contamination by other radionuclides, and therefore it is essential to hold this contamination to as low a level as is possible. A contaminating radionuclide does not add to diagnostic information, but it does increase the radiation dose to the patient and, in many cases, may degrade the quality of the scan. A good example is provided by the radionuclide, $^{123}$I which is difficult to produce without $^{124}$I as radiocontaminant. Iodine-124, besides significantly increasing radiation dose to the patient, degrades image quality because of its emission of high energy gamma rays.

The amount of impurity is generally given as $\mu$Ci of radiocontaminant per microcurie or millicurie of the desired radionuclide. Sometimes the limit of the allowable contamination is set by governmental agencies, as in the case of $^{99m}$Tc, for which the amount of $^{99}$Mo must not exceed more than 0.15

$\mu$Ci for each mCi of $^{99m}$Tc. In cases where no such limits are prescribed, the rule of thumb is to keep the radiation dose to the patient from the radiocontaminants to less than 10% of that due to the radionuclide of interest.

Another important aspect of radionuclidic purity is that it does not stay constant with time. Where the half-life of the desired radionuclide is shorter than that of the radiocontaminant, radionuclidic purity degrades with time and vice versa. For example, since the half-life of $^{124}$I, a common radiocontaminant in $^{123}$I, is longer than $^{123}$I, the radionuclidic purity is best at the time of the production of this radionuclide, and as the radionuclide is stored, it becomes progressively less pure.

The most common method of determining the nature and extent of radionuclidic impurity is with gamma spectroscopy using a NaI(Tl) or Ge(Li) detector both of which are discussed in Chapter 8.

***Radiochemical Purity:*** Since a radionuclide may form several compounds with a given chemical, it is important to ascertain that a given radiopharmaceutical is in the desired chemical form. Any radiochemical impurities present should be precisely stated. In this regard, it is also important to consider that although a radiochemical may be pure to begin with, it may not be stable over a period of time as a result of the action of radiation or the nature of the chemical itself. To avoid this deterioration, the radiochemical should be stored properly according to the instructions of the manufacturer. For example, radioiodinated human serum albumin (RIHSA), which, among other things, is used as a blood pool scanning agent, may be 99.9% pure when freshly prepared. With time, however, some of the radioiodine becomes free. The amount of free radioiodine strongly depends upon storage conditions. The contamination with free radioiodine is several times higher if RIHSA is stored at room temperature than if it is refrigerated. A significant amount of free radioiodine will interfere with the intended study.

A common method for the detection of radiochemical impurities is thin layer or paper chromatography.

***Chemical Purity:*** A radiopharmaceutical should contain only the desired chemical. In the final preparation of the radiopharmaceutical there may be a number of chemicals involved besides the radiochemical of interest. All these chemicals should be compatible with each other in vitro as well as safe for the patient. In addition, these must not distort the in-vivo function of the main chemical.

***Sterility:*** A radiopharmaceutical should be sterile, *i.e.*, free from any microbial contamination, and, therefore, should be tested to this effect before use in patients. In the case of radiopharmaceuticals using short-lived radionuclides ($^{99m}$Tc and $^{113m}$In), where prior testing of the final product is not feasible, the sterility of the labeling technique should be tested adequately and periodically.

***Apyrogenicity:*** Even if the preparation is sterile it may still contain pyrogens which, when intravenously administered to a patient, may cause a reaction. A radiopharmaceutical, therefore, should also be tested for pyrogenicity before use in humans. If this is not feasible, as in the case of short-lived radionuclides, the apyrogenicity of the technique should be ascertained properly and periodically.

## TECHNETIUM 99m-LABELED RADIOPHARMACEUTICALS

Because of the very attractive physical characteristics of $^{99m}Tc$, a variety of chemicals have been labeled with this radionuclide in spite of the fact that the exact mechanism of technetium-labeling with these compounds is often not known. Technetium 99m, in the form of sodium pertechnetate ($Na^{99m}TcO_4$), is easily obtained in the laboratory from a $^{99}Mo$-$^{99m}Tc$ generator. The labeling of the majority of chemicals by $^{99m}Tc$ is achieved by first reducing the pertechnetate to ionic technetium (mostly $Tc4+$) and then complexing it with the desired chemical. The common agent used for reducing purposes is stannous chloride ($SnCl_2$). Since the half-life of $^{99m}Tc$ is short (6 hr), most of the labeling has to be performed "in-house." This is greatly simplified by the use of sterile and pyrogen-free kits, in which all the desired chemicals are premixed and held together in a lyophilized state under an inert atmosphere (nitrogen gas). To label a particular chemical compound, it is necessary only to introduce a known amount of sterile and pyrogen-free sodium pertechnetate into the kit vial; the labeled compound is ready to use within a few minutes.

Three parameters—labeling efficiency, in-vitro stability and in-vivo stability—are important considerations in the selection of a kit. Labeling efficiency is defined as the % of total radioactivity present in the kit which is tagged to the appropriate molecule or compound. For most kits currently in use in nuclear medicine, labeling efficiencies under optimum conditions are in excess of 90%, sometimes even reaching as high as 99%. The remainder of the radioactivity (which is not tagged to the desired compound) is present as radiochemical impurity. In kits which use $SnCl_2$ as the reducing agent, radiochemical impurities are, in general, of two forms—free pertechnetate (which was not reduced) and reduced or hydrolyzed technetium (which was reduced but did not tag to the compound of interest). Reduced or hydrolyzed technetium sometimes forms a colloid with excess tin present in the kit. This is also a radiochemical impurity but in a different form.

The in-vitro stability of a labeled compound determines the time it can be stored on the shelf without significant deterioration. A high in-vitro stability allows the compound to be labeled once and then be used in a number of patients at different times of the day. Both labeling efficiency and in-vitro stability of compounds prepared from kits can be maximized by observing a few simple precautions such as using only oxident free sodium pertechnetate solution and making sure no air is introduced into the reaction vial during the labeling procedure. The in-vitro stability of a kit can also be extended by the use of preservatives in the reaction vial as is done by some manufacturers and/or by storing the labeled material at low temperatures. I should caution here that in-vitro stability of a labeled compound is different from in-vitro stability of the kit or the chemical compound itself.

The in-vivo stability of a labeled compound determines how closely the distribution of the radiolabeled compound in the biological system parallels that of the unlabeled compound. The distribution of a labeled compound should be similar to the unlabeled compound at least for the duration of the study.

The distribution and use of the most common technetium-labeled compounds in nuclear medicine are described below. Most of these compounds can be easily and rapidly prepared from the commercially available kits.

***Technetium 99m Pertechnetate ($^{99m}TcO_4^-$):*** This radiopharmaceutical is obtained directly from the $^{99}$Mo-$^{99m}$Tc generator using saline as the eluting solution. In biological systems it behaves similarly to iodine. After oral or intravenous administration it is selectively concentrated in the thyroid, salivary glands, stomach and choroid plexus. Disappearance of the pertechnetate from plasma is a multiexponential function. About 50% of the compound is rapidly diluted into extravascular spaces within 15 to 20 min. The remaining amount disappears from the plasma with a half-life of about 3 hr. About 20 to 30% of the injected dose is excreted eventually in feces at a slow rate.

Stomach, which is the major organ of uptake, contains 20 to 25% of the injected dose at 4 hours. So much radioactivity remains in the stomach even after 24 hours that it is not advisable to perform scanning of an abdominal organ using $^{99m}$Tc labeled radiopharmaceuticals (or whose principal gamma emission is 170 keV or below) on a patient, who had a brain scan up to 48 hours prior to the intended study.

Technetium 99m pertechnetate is, at present, the agent of choice for brain scanning. It is also used for thyroid, salivary gland and stomach scanning.

***Technetium 99m-Labeled Sulfur Colloid:*** This radiopharmaceutical is easily prepared using commercially available kits. Colloids in general are removed from the blood stream by the reticuloendothelial (RE) cells of the body. The relative distribution of the colloids among the RE cells of the various organs depends on factors such as the size, nature and amount of colloidal particles, blood supply to the organ, and on other physiologic and pathophysiologic considerations. In the case of colloidal sulfur tagged with technetium 99m (particle size ~0.3 $\mu$ where $\mu = 10^{-4}$ cm), about 70 to 80% of the injected dose is localized in the liver within 10 to 20 min of intravenous administration. Of the remaining amount, about 3% is deposited in the spleen and about 15 to 20% is localized in the bone marrow. This agent is, therefore, primarily used for liver, spleen and bone-marrow scanning.

Other $^{99m}$Tc labeled compounds sometimes used for liver, spleen and bone marrow scanning are antimony sulphide colloid, stannous hydroxide colloid and albumin micro-aggregates.

***Technetium 99m-Labeled Macroaggregated Albumin ($^{99m}$Tc MAA):*** This radiopharmaceutical is primarily used in lung scanning. Within a few seconds after intravenous administration of $^{99m}$Tc MAA, 90 to 95% of the injected dose is trapped in the capillary and precapillary bed of the lungs. For effective lung localization, the albumin macroaggregates must be between 15 to 75$\mu$ in size. The biological half-life of $^{99m}$Tc MAA in the lungs is about 8 to 12 hr.

***Technetium 99m-Labeled Polyphosphate, Pyrophosphate and Diphosphonate:*** These radiopharmaceuticals are primarily used for bone scanning. After an intravenous injection, about 50 to 60% of the injected dose is localized in the skeleton within 15 to 20 min. The remainder of the dose is distributed in soft tissue and plasma from which it is excreted slowly in the urine. About 20 to 30% of the injected dose is either excreted or taken up by the kidneys within 3 hr of the injection. Of these three groups, polyphosphate has the slowest plasma clearance, and is, therefore, least desirable for bone scanning. Methylene disphosphonate (MDP) has the fastest plasma clearance of these compounds.

Use of $^{99m}$Tc pyrophosphate and $^{99m}$Tc diphosphonate for the detection of myocardial infarctions is now well established.

***Technetium 99m-Labeled Human Serum Albumin:*** This radiopharmaceutical is primarily used for blood pool scanning as in the case of the heart or placenta. After intravenous administration it is retained in the plasma for a long period of time. However, $^{99m}$Tc-labeled albumin is not as stable in vivo as albumin labeled with radioiodine or radiochromium. Therefore it is not the preferred agent for plasma volume determination.

***Technetium 99m-Labeled Red Cells:*** Labeling of red cells with radionuclides, in general, is a complex and time-consuming procedure. Recently, however, a simple and expeditious method has been developed to label red cells in vivo with $^{99m}$Tc. This has increased the popularity of $^{99m}$Tc labeled red cells over that of $^{99m}$Tc labeled albumin for blood pool scanning, particularly for heart. In this method two steps are involved. In step one, a patient's red cells are coated with stannous ion ($Sn^{++}$). This is achieved simply by injecting about $\frac{1}{5}$ of a "cold" pyrophosphate kit into the patient. A "cold" kit is a preparation reconstituted using saline only, *i.e.* without any radioactivity. In step two which is performed about 30 minutes later, the desired amount of the $^{99m}$Tc radioactivity in the form of pertechnetate is administered to the patient as a second injection. The "pretinned" red cells in the patient quickly reduce the pertechnetate to ionic technetium which then readily binds to the red cell surface. Little free pertechnetate or reduced technetium remains in circulation. Most of the radioactivity (90%) is bound to the patient's red cells.

The second step—tagging of $^{99m}$Tc to the pretinned red cells—can also be performed in vitro by withdrawing pretinned red cells from the patient and incubating them with the desired amount of $^{99m}Tc\ O_4^-$. The red cells thus labeled may be heat damaged, if desired, by heating them in a water bath at 50°C for half an hour. Damaged red cells are used to image the spleen, without interference from the liver which is the case when sulfur or other colloids are used to image the spleen.

***Technetium 99m-Labeled 2,3-Dimercapto Succinic Acid (DMSA):*** This radiopharmaceutical is the agent of choice when morphology of the renal cortex is of interest. After an intravenous administration, this radiopharmaceutical is quickly mixed with the plasma volume from where it is cleared with a half time of about 1 hour. By 2 hours, between 40 and 50% of the injected dose is taken up by the renal cortex and about 15% is excreted in urine. Because of rapid in-vitro decomposition of $^{99m}$Tc labeled DMSA, it should be stored in a refrigerator and used within half an hour of labeling.

***Technetium 99m-Labeled Diethyltriamine Pentaacetic Acid (DTPA):*** This radiopharmaceutical is used primarily for brain and kidney scanning. After intravenous administration, $^{99m}$Tc DTPA is rapidly cleared by the kidneys. The biological half-life in plasma of DTPA chelates in man is about 15 min. Over 80% of the injected dose can be recovered in urine between 2 to 3 hr postinjection.

***Technetium 99m-Labeled Glucoheptonate:*** Biological behavior of this radiopharmaceutical is somewhere between that of ($^{99m}$Tc) DTPA and ($^{99m}$Tc) DMSA. Its clearance from plasma is slower than that of $^{99m}$Tc DTPA but faster than $^{99m}$Tc DMSA. Maximum uptake by renal cortex, which is reached at about 1 hour postinjection, is about 25% of the injected dose. Another 25% of the injected dose is excreted in urine by this time. Besides renal scanning, this radiopharmaceutical is quite often used for brain scanning. There is some evidence that this radiopharmaceutical may be better than $^{99m}$Tc $O_4^-$ for brain scanning. Another use of this radiopharmaceutical is in detection of myocardial infarctions at an early stage (within 2 to 3 days postinfarction). At this early stage $^{99m}$Tc pyrophosphate is not useful for the detection of myocardial infarctions.

***Technetium 99m-Labeled Mertiatide (Mag3):*** This compound has recently been introduced to study renal function and serves as a substitute for iodohippuran. Following intravenous injection, this compound is rapidly cleared by the kidneys through both active tubular secretion and glomerular filtration. In normal subjects, 90% of the injected dose can be recovered in urine. This compound, even though bound to plasma proteins, has a fast plasma clearance because of the reversible nature of the binding.

***Technetium 99m-Labeled 2,6-Dimethyl Acetanilide Iminodiacetic Acid (better known as HIDA) and Related Compounds (Diethyl-IDA, PIPIDA and DISIDA):*** These radiopharmaceuticals are used for imaging the hepatobiliary system and now have replaced the use of $^{131}$I rose bengal for such purpose. In normal subjects, $^{99m}$Tc HIDA is rapidly cleared from blood by hepatocytes. The half time for blood clearance is approximately several minutes. The transit of the radioactivity from the liver through gall bladder to the intestine is also fast. Within an hour after administration, over 70% of the injected dose is in intestine. About 15% of the injected dose is excreted in urine during the first hour. The remainder of the activity is eventually excreted in feces. Urinary excretion of diethyl-IDA is smaller than that of PIPIDA or DISIDA. The transit and excretion of these compounds through the hepatobiliary system is strongly dependent on the patency of this system.

***Technetium 99m-Labeled Sestamibi (Cardiolite):*** This compound has shown great promise in the imaging of the heart and is a good candidate to replace thallium-201 in clinical nuclear medicine. Being a technetium-labeled compound, it has an advantage over thallium in terms of providing higher photon flux for the same radiation dose to the patient. After intravenous injection, it is extracted by the myocardium in proportion to the blood flow. Its first-pass extraction efficiency is slightly less than that for thallous ion, but it does not redistribute and remains in the myocardium up to 3 hours post-injection, thus providing a long time for planar imaging as well as SPECT. The liver and kidneys are other organs of large localization (20 and 14% respectively). It is excreted intact through the hepatobiliary system and the kidneys.

***Technetium 99m-Labeled Teboroxime (CardioTec):*** This is another technetium-labeled compound for imaging the myocardium. After an intravenous injection, it is rapidly cleared from the blood. The peak activity in the myocardium is achieved within a few minutes, after which it starts to redistribute. This fact is its strength as well as its weakness: strength because it allows rest and stress studies to be performed in less than an hour, a marked improvement over Sestamibi; and weakness because there is not enough time to get good quality planar images and almost no time to perform SPECT. The liver is the organ with maximum localization, 33% at 15 minutes post-injection. The primary route of excretion is the hepatobiliary system, with some excretion through the kidneys.

***Technetium 99m-Labeled Brain Imaging Agents [Exametazime (Ceretec), Hexamethylpropylene amine oxime (HMPAO), and ethyl cysteinate dimer (ECD)]:*** These radiopharmaceuticals cross the blood-brain barrier and therefore actively localize in the brain. Since their uptake is generally a function of regional blood flow, these agents, in conjunction with SPECT, are now being used to measure brain function. After an intravenous injection of Ceretec, the peak uptake (7%) in the brain is reached at 1 minute. The remainder of the dose is distributed in the whole body, particularly in muscle and soft tissues. It is excreted from the body by way of the GI tract and the kidneys in similar amounts.

## RADIOIODINE-LABELED COMPOUNDS ($^{131}$I and $^{123}$I)

With the advent of technetium 99m-labeled compounds, use of iodine 131-labeled compounds for scanning purposes has declined sharply. However, two radioisotopes of iodine, $^{123}$I and $^{131}$I, are commonly used for diagnosis of thyroid diseases. $^{131}$I is also used in the treatment of hyperthyroidism and thyroid cancer. For diagnostic purposes, $^{123}$I (13-hour half-life and 87% 160-keV $\gamma$-ray emission) is slowly emerging as the radionuclide of choice because it delivers significantly less radiation dose to the thyroid than $^{131}$I. The two disadvantages of $^{123}$I which have slowed its widespread use are its cost and, sometimes, the presence of large amounts of other radioisotopes of iodine such as $^{124}$I. These contaminants not only increase radiation dose, but also degrade image quality. Both $^{131}$I and $^{123}$I are commercially available in capsules or solutions with high specific activities almost carrier-free. Since iodine is readily absorbed from the stomach, it is generally administered orally.

After an intravenous administration, the iodide ion is very quickly distributed throughout the extracellular water. From there it is slowly taken up by the thyroid, stomach and intestine, salivary glands and choroid plexus. A large fraction is filtered into the urine by the kidneys. By 24 hours, about

75% of the injected dose is excreted, 15% taken up by thyroid, 4 to 5% is in the GI tract, and 1 to 2% is circulating in the blood. In the thyroid, iodine is then organified and produces thyroid hormones such as $T_3$ or $T_4$. The distribution of radioiodide can be drastically changed in some disease states, particularly hyperthyroidism or severe kidney malfunctions. In hyperthyroidism, thyroid can take up to 90% of the administered dose with practically no excretion, whereas in hypothyroidism, thyroid takes up little (sometimes less than 1 to 2%) iodide with excretion increased up to 95%.

Two other iodine-123 labeled compounds, $^{123}$I-Hippuran and $^{123}$I-isopropylamphetamine (IMP or related compounds) are finding increased use in nuclear medicine. I-123-Hippuran is rapidly cleared by the kidneys and therefore is used to study renal function in the form of renograms which can be obtained either using dual scintillation detector probes or by imaging with a scintillation camera interfaced with a computer. I-123 isopropylamphetamine is used for measuring brain function—particularly the changes in regional brain blood flow. Brain function studies with $^{123}$I amphetamines in conjunction with single photon emission tomography are promising as an alternative to the use of $^{18}$F deoxy glucose in conjunction with positron emission tomography which necessitates an in-house cyclotron and associated paraphernalia.

## COMPOUNDS LABELED WITH OTHER RADIONUCLIDES

***Gallium-67 Citrate:*** This radiopharmaceutical is employed in detection of soft tissue tumors and inflammatory diseases. After an intravenous administration, a significant fraction of the gallium in blood (30%) is bound to plasma proteins—in particular to transferrin. The remainder of the gallium quickly diffuses into extracellular spaces and is slowly cleared by the kidneys. By 24 hours, about 15% of the dose is excreted in urine and about 10% is circulating in the blood. The remainder of the radioactivity is distributed in kidneys, bone, liver and lymph nodes. Biological half-life of gallium in man is between 1 and 2 weeks. Therefore, significant amounts of this radionuclide persist in body even after 2 weeks. Besides excretion in urine (about 25% in 1 week), there is significant (10%) excretion in stool which sometimes interferes in interpretation of scans of the abdominal area. The exact mechanism of localization of $^{67}$Ga in tumors or inflammatory lesions is not well established at this time.

***Thallous-201 Chloride:*** This radiopharmaceutical is primarily used for detection of myocardial infarction and/or ischemia. Thallous ion mimics the biological behavior of potassium which is avidly localized intracellularly. Like potassium, immediately after intravenous injection thallous ion quickly leaves the circulation ($T\frac{1}{2}$ = 4 min) and is taken up by various organs generally in proportion to the blood supply of the given organ (brain is one exception where there is almost no accumulation). At about 15 to 20 minutes, 4% of the injected dose is localized in myocardium, 12% in liver, 4% in kidneys and the bulk of the remainder is distributed in muscles throughout the body. Biological half-life of thallous ions in man is about 10 days. Its use for myocardial imaging is based on the fact that its cellular uptake is dependent on blood flow to that region and the integrity of the cells themselves. Therefore, reduced uptake of thallous ion in a region is either a reflection of reduced blood flow (ischemia) or a reflection of damage to the cells (infarction).

***Chromium-51 Labeled Red Cells:*** These are used to determine the red cell volume and red cell mean life. Since labeling of red cells has to be performed in-house for each patient, this is a complex procedure. First, blood is withdrawn from the patient. It is then incubated with acid dextrose (ACD) and sodium chromate-51 solution at 37 to 39°C temperature for 10 minutes. Chromate ion readily penetrates the red cell membrane. Inside the cell it is reduced to $Cr^{+++}$ which has high affinity for hemoglobin. Small amounts of chromate ion which do not enter the red cells are then reduced to $Cr^{+++}$ outside the cells by addition of appropriate amount of ascorbic acid. The blood-ACD-ascorbic acid solution is now ready to be injected back into the patient. Only the labeled red cells remain in circulation. Small amounts of $^{51}Cr^{+++}$ not bound to the cells are quickly excreted in urine by the kidneys.

***Indium-111 Labeled DTPA:*** This radiopharmaceutical is presently the agent of choice for imaging cerebrospinal fluid (CSF) dynamics. When injected intrathecally, it rises to the basal cisterns within 2 to 3 hours. By 24 to 48 hours, radioactivity rises over the convexities of the brain and into the parasagittal area. There, it is resorbed into blood and quickly excreted by the kidneys into urine.

***Indium-111 Labeled Platelets and Leukocytes:*** These are used for thrombus and abscess detection respectively. The technique of platelet or leukocyte labeling with radionuclides is quite involved as it first requires separation of platelets or leukocytes from other components of blood. Once platelets or leukocytes have been separated, they can be labeled with $^{111}In$ by incubating them with $^{111}In$ oxine at room temperature for 30 minutes. Because of the involved preparation of $^{111}In$ labeled platelets or leukocytes, these have not found widespread use in nuclear medicine so far.

***Indium-111 Labeled Monoclonal Antibodies:*** Wide interest exists in the possible use of monoclonal antibodies labeled with different radionuclides for diagnosis and therapy of cancer and its metastases. Because of the affinity of tumor antibodies for tumor, a high degree of localization occurs in tumors when antibodies labeled with radionuclides are injected in patients with a tumor. This localization can be used for diagnosis by labeling antibodies with radionuclide such as $^{111}In$ or for therapy with $^{131}I$. Initial results are encouraging and work is in progress to define the clinical potential of these radiopharmaceuticals.

## RADIOACTIVE GASES

Of all the radioactive gases used in nuclear medicine $^{133}Xe$ has found the most popularity because of its easy availability and useful physiologic properties. It can be used in a gaseous form or as a saline solution. In a gaseous form it is used for lung ventilation studies; in a saline solution administered intravenously, it is used for lung perfusion studies. When $^{133}$Xe is injected intra-arterially in a saline solution, it can be used to measure the blood flow to the organ supplied by that artery.

Biological half-life of xenon in the body is only a few minutes. Only a very small component (~2%), probably the portion which is transferred to fat in the body, has a much longer biological half-life (~10 hr).

Recently, with the commercial availability of $^{81}$Rb − $^{81m}$Kr generators, many institutions have started using $^{81m}$Kr instead of $^{133}$Xe. Because of 13 seconds half-life of $^{81m}$Kr, repeat studies can be performed in the same patient within a few minutes of the first study. Also, use of $^{99m}$Tc DTPA aerosols for such purpose is under progress. Initial results seem quite promising.

## THERAPEUTIC USES OF RADIOPHARMACEUTICALS

The main concern of nuclear medicine is diagnostic applications of radionuclides. Here, we shall discuss briefly some of the therapeutic uses of radiopharmaceuticals when they are administered internally to a patient.

***Design of a Radiopharmaceutical for Therapeutic Uses:*** Physiologic and biochemical considerations in designing a radiopharmaceutical for therapeutic purposes are similar to those which apply in designing a radiopharmaceutical for diagnostic uses: the primary objective in both cases is a high target to non-target ratio. For therapeutic uses, a diseased cell is the target.

Since the objective of therapeutic use of a radiopharmaceutical is to kill diseased cells with radiation, the physical requirements for a radionuclide are the opposite of those desired for diagnostic purposes. The radionuclide should emit only particulate radiation ($\alpha$ or $\beta$) and preferably no penetrating radiation (x or $\gamma$).

***Problems and Uses:*** The main problem arising in the treatment of a patient with internally administered radionuclides is the calculation of the radiation dose (rads) to the target. Accurate calculations of the radiation dose require accurate data about the physical decay characteristics of the radionuclide and its distribution in the body (see Chapter 7). Although accurate data are available with regard to the first parameter, accurate data with regard to the latter are difficult to obtain, particularly in the case of a patient who is about to be treated. Usually, a tracer dose is administered to a patient for this purpose and from the blood-disappearance, and the urinary and fecal excretion rates of this tracer dose, as well as its relative distribution in the body as determined by area scanning where possible, an estimation is made of the various parameters to be used in the calculation of radiation dose. Unfortunately, even these parameters may not describe the behavior of the therapy dose (100 to 1,000 times larger than the tracer dose) and, therefore, the resulting radiation dose from a therapy dose may be quite different from that estimated. Another unpredictable factor in the response of a patient to such treatment is the variability of the biological response of different individuals to the same radiation dose.

**Table 5–2. Therapeutic Uses of Radiopharmaceuticals**

| *Radionuclide and Chemical Form* | *mCi, and Route of Administration* | *MBq* | *Uses* |
|---|---|---|---|
| $^{131}I$ as iodide | 3–10 mCi, oral | 111–370 | Hyperthyroidism |
| $^{131}I$ as iodide | 50–200 mCi, oral | 1850–7400 | Cancer of thyroid |
| $^{32}P$ as phosphate | 3–20 mCi, intravenous | 111–740 | Polycythemia, bone metastases, and leukemia |
| $^{32}P$, $^{198}Au$, $^{90}Y$, $^{177}Lu$ in colloidal form, particle size ranging from .009 to 50$\mu$ | 10–150 mCi, intravascular, intracavitary and intrastitial | 370–5550 | A variety of malignant diseases |

In spite of this inability to deliver the exact number of rads to a target by internal administration of the radionuclides, success has been achieved for the treatment of a number of diseases. Table 5–2 summarizes the present status of the therapeutic uses of internally administered radionuclides.

## PROBLEMS

1. Three radionuclides with the following properties are available to you: (a) $T_{\frac{1}{2}}$ = 3 minute, $\gamma$ emission (70%) of 180 keV energy, (b) $T_{\frac{1}{2}}$ = 1 day, $\gamma$ emission (100%) of 250 keV energy, a small fraction of conversion electrons (5%), (c) $T_{\frac{1}{2}}$ = 2 days, $\gamma$ emission (20%) of 300 keV, $\beta$ emission with a maximum energy of 2.0 MeV. Which one of these is best suited for (a) tumor localization when the optimum tumor-uptake occurs at 18 hours postinjection, (b) blood perfusion in an organ, and (c) radiation therapy?
2. List the factors that affect the localization of a radiopharmaceutical in tissue.
3. What happens to the radionuclidic purity with time if the half-life of the desired radionuclide is longer than that of the radiocontaminant?
4. How does the presence of oxygen interfere with labeling of a pharmaceutical with $^{99m}Tc$?
5. List the various radiopharmaceuticals that are used for imaging the following organs: liver, bone, bone marrow, myocardium, blood pool, kidneys, lungs, hepatobiliary system, thyroid, spleen, tumors, and brain.

# 6

# Interaction of High-Energy Radiation with Matter

Interaction is a fundamental aspect of nature. Our ability to see, hear, smell, and taste is a vivid manifestation of interaction. This chapter, however, focuses on the mechanism of interaction of high-energy radiation with matter. Radiation in this context is used in a general sense. It encompasses both corpuscular radiation (*e.g.*, charged particles and neutrons) and electromagnetic radiation in the form of x- and $\gamma$-rays.

The material presented here is basic to an understanding of the detection and effects (especially biological) of high-energy radiation as well as of protection against it. Since the subject matter is complex, only the salient features are presented. For ease of presentation and understanding, this chapter has been divided into three sections: (1) interaction of charged particles (with energies of 10 keV to 10 MeV) such as e, $e^+$, p, $\alpha$, and $^2D$; (2) interaction of high-energy photons such as x- and $\gamma$-rays; and (3) interaction of neutrons.

## INTERACTION OF CHARGED PARTICLES (10 keV TO 10 MeV)

### Principal Mechanism of Interaction

When a charged particle passes through a substance (target) it interacts with the negatively charged electrons and positively charged nuclei of the target atoms or molecules. Through the coulomb forces, it tries to attract or repel the electrons or nuclei near its trajectory. As a result of these pushes and pulls (a sophisticated name for these is inelastic collisions) the charged particle loses some of its energy which is taken up by the electrons of the target atoms near its trajectory. The absorption of energy by the target atom leads to its ionization or excitation. In this energy range (10 keV to 10 MeV) ionization events predominate over excitation events. For this reason high-energy radiations are sometimes referred to as ionizing radiations, although excitation events are by no means negligible. The probability of inelastic collisions in general is so high that it does not take a material of much thickness to stop the charged particles completely.

## Differences Between Lighter and Heavier Charged Particles

Do all charged particles interact in a similar way? The answer is yes and no. Yes, because inherently the nature of interaction for all charged particles in this energy range is the same (inelastic collisions). No, because the manifestation of these interactions on lighter particles—whose masses are of the order of an electron (*e.g.*, e and $e^+$)—and heavier particles—whose masses are equal to or more than that of a proton (*e.g.*, p and $\alpha$)—is strikingly different. Lighter particles in inelastic collisions with the electrons of the target atoms, besides losing energy, tend to be deflected at larger angles than the heavier particles. This leads to a wide variation in the paths of the two kinds of particles (depicted graphically in Fig. 6–1 for an electron and proton). The path of a heavier particle is more or less a straight line, whereas that of a lighter particle is more tortuous (zigzag). Whenever a lighter charged particle is deflected at a large angle, the energy transferred to the target electron is also quite large. As a result, the target electron acquiring this large amount of energy also behaves like a high-energy charged particle, thus creating its own path in the target medium. The paths created by high-energy secondary electrons are known as $\delta$ rays. These are shown by the dotted lines in Fig. 6–1. In the case of a proton or other heavier particles, this type of energy transfer is rare.

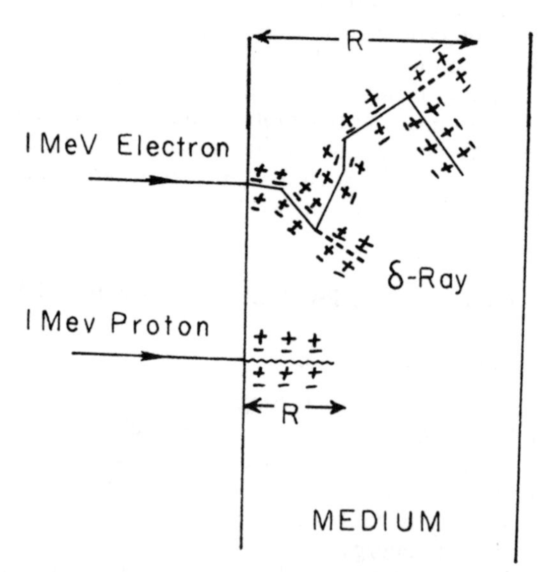


**Fig. 6–1.** Range, R, of a charged particle. Heavy charged particles (e.g., protons) travel and lose their energy in more or less straight lines. Light charged particles (e.g., electrons) lose their energy in a zigzag fashion. Lighter charged particles can transfer large amounts of energy in a single encounter with an electron of the medium, thus creating what are known as $\delta$-rays. Plus and minus signs represent the ionization of the atoms of the medium. (Fig. not to the scale as electrons will travel much farther than protons.)

**Table 6–1. Approximate Ranges of Charged Particles**

| | Range R (cm) | | | |
|---|---|---|---|---|
| | Soft Tissue | | Air | |
| *Energy (keV)* | *e or* $e^+$ | $\alpha$ | *e or* $e^+$ | $\alpha$ |
| 10 | $2 \times 10^{-4}$ | $<10^{-5}$ | $1.6 \times 10^{-1}$ | $1 \times 10^{-2}$ |
| 100 | $2 \times 10^{-2}$ | $1.4 \times 10^{-4}$ | 16 | $1 \times 10^{-1}$ |
| 1,000 | $4 \times 10^{-1}$ | $7.2 \times 10^{-4}$ | $3.3 \times 10^{2}$ | $5 \times 10^{-1}$ |
| 10,000 | 5 | $1.4 \times 10^{-2}$ | $4.1 \times 10^{3}$ | 10.5 |

## Range, R, of a Charged Particle

As a charged particle travels farther and farther in a medium, it loses more and more energy and, therefore, more and more of the target atoms close to its path become ionized or excited. Eventually the charged particle loses all its kinetic energy and comes almost to a halt. The average distance traveled by a charged particle in the incident direction is defined as its range, R. This definition of range R is strictly valid only for heavy charged particles such as $\alpha$ particles. For lighter particles, it is difficult to define the range R exactly. For our purpose, it is sufficient to think of the range R of light particles such as electrons or positrons as the minimum thickness of a material which they are just unable to penetrate. The concept of the range of a charged particle is quite useful in radiation protection, design of radiation detectors and radiation dosimetry.

The range, R, of the heavier charged particles, since they travel more or less in straight lines, is nearly equal to the average path length of the charged particles in a given medium, whereas the range, R, for the lighter particles such as electrons, because of their more tortuous paths, is much shorter than the average path length (see Fig. 6–1). Table 6–1 lists the ranges of $\alpha$ particles and electrons for various energies and mediums, providing a rough idea of the distances involved.

The concept of range is sometimes used with $\beta$-emitting radionuclides in which $\beta$ particles are emitted with varying amounts of energy up to a maximum, $E_\beta$ max. In this case the range is more or less determined by $E_\beta$ max.

## Factors Which Affect Range, R

The range, R, of a charged particle depends on various factors. Four of the most important of these are described below.

***Energy (E):*** The range, R, of a given particle increases with increase in the initial energy, E, of the particle. For example, the range of a 5-MeV electron is about 6 times longer than that of a 1-MeV electron. The exact relationship of R to E is complex, but in the energy range of our interest, R is linearly related to the initial energy of the charged particle E or $R = AE + B$ where A and B are constants.

*Mass (M):* Lighter particles have longer ranges than heavier particles of the same energy and charge. The range of a 1-MeV positron ($e^+$) is much longer than the range of a 1-MeV proton ($p^+$, 2000 times the mass of an $e^+$). Mass dependence of the range is sometimes expressed as velocity dependence. A 1-MeV positron is traveling at a much higher speed than a 1-MeV proton. The range, R, of a charged particle increases as the velocity of the charged particle increases.

*Charge (Q):* A particle with less charge travels farther than a particle with more charge. For example, a $^3_1H$ (charge 1; mass 3) particle has a longer range than a $^3_2He$ (charge 2; mass 3) particle of the same energy. The exact relationship is $R \alpha \frac{1}{Q^2}$. Therefore, $^3_1H$ particle travels four times the distance of $^3_2He$ particles. The signs of the charge (positive or negative) do not affect the range.

*Density of the Medium (d):* The range, R, of a charged particle strongly depends upon the density of the medium through which it is traversing. The higher the density of the medium, the shorter the range of a charged particle, *i.e.*, R is inversely proportional to the density, d, of the medium or $R \alpha \frac{1}{d}$. For this reason the ranges of charged particles are always much longer in gases than in liquids or solids.

### Bremsstrahlung Production

Besides losing energy through inelastic collisions, charged particles can lose energy through the "bremsstrahlung" process. In this case, when a charged particle in the electric field of a nucleus experiences a sudden acceleration or de-acceleration, it sometimes emits high-energy photons (x-rays). The probability of such interaction in the energy range of our interest is indeed very small except when electrons and positrons interact with high atomic-number materials (*e.g.*, lead, steel). The fraction of energy, f, released as x-rays by electrons of energy E (MeV) in a material with atomic number Z is given approximately by the following equation:

$$f = \frac{Z \cdot E}{1400} \text{ (E is in MeV)}$$

In an x-ray tube with a tungsten target (Z = 74) and electrons with 100-keV incident energy, only 0.5% of the total energy of the electrons is converted into x-rays. The same fraction (0.5%) of the total energy of the electrons is converted into x-rays when a 1-MeV electron passes through water (Z = 7.4) except that the x-ray spectrum (energies of emitted x-rays) in this case is quite different from that produced in the first case (x-ray tube). For the purposes of nuclear medicine, since production of high-energy photons by this mechanism is generally very small, it is therefore ignored.

### Stopping Power, S

Quite often, instead of using the range, R, of a charged particle, another parameter known as stopping power, S, is used. Stopping power is defined as the ratio of the amount of energy lost (dE) by a charged particle in traversing a small distance (dx) in a given medium to the distance dx, or

$$S_{medium} = -\frac{dE}{dx}$$

The range, R, is related to the stopping power in an approximately inverse relationship, *i.e.*, the higher the stopping power of a given medium, the shorter will be the range of a given particle in that medium. Because of this relationship, the stopping power, S, also depends on the same factors (E, M, Q and d) on which the range, R, depends, although now the relationships are approximately inverse to that of the range, R.

### Linear Energy Transfer (LET)

Linear energy transfer is an important parameter in radiation biology. It is defined as the ratio of the amount of energy transferred ($dE_{local}$) by a charged particle to the target atoms in the immediate vicinity of its path in traversing a small distance (dx) to the distance dx, or

$$LET = -\frac{dE_{local}}{dx}$$

### Difference Between LET and Stopping Power, S

In discussing stopping power, S, one is concerned with the total energy lost by the particle, whereas in discussing LET one is concerned with the local (immediate vicinity of the track) deposition of energy by the charged particle. Because of this, S and LET are almost equal for heavier particles. For lighter particles, however, the two quantities differ significantly because of the energy lost in the $\delta$ rays or through the bremsstrahlung process which does not deposit energy locally. Both the stopping power, S, and LET are measured in the same units (keV/$\mu$, $1\mu = 10^{-4}$ cm).

### Annihilation of Positrons

There is no difference in the way in which an electron or a positron loses energy in a medium. However, a positron, once it has lost its energy, is not stable and is quickly annihilated by combining with an electron. The mass energy of the electron and the positron is converted into two $\gamma$ rays of 511 keV which travel in opposite directions. Remember that the annihilation of positron occurs only when it has lost almost all of its energy, *i.e.*, near the end of its range in a medium.

## INTERACTION OF X- OR $\gamma$-RAYS (10 keV TO 10 MeV)

Although a photon does not have any electrical charge, it nevertheless interacts with electrical charges and, therefore, with matter which is composed of electric charges. The probability of interaction and the modes of interaction of a photon with matter are strongly dependent on the energy of the photon. Here we are interested only in the interaction of high-energy photons (x- or $\gamma$-rays) with matter.

The probability of interaction of an x-ray or $\gamma$-ray with an atom is, in general, very small compared to that of high-energy charged particles. As a consequence, x- or $\gamma$-rays have more penetrating power than high-energy charged particles. Because of this high penetrating power of x- or $\gamma$-rays, it is not practical to use concepts such as stopping power or range. Instead, new concepts known as linear attenuation coefficient ($\mu$ [linear]) and the half-value layer (HVL) are commonly used.

## Attenuation and Transmission of X- or $\gamma$-Rays

Let us consider a simple experiment in which a parallel beam of x- or $\gamma$-rays of a given energy ($E_\gamma$) is incident on a thin slab of 1-$cm^2$ cross section and x-cm thickness (Fig. 6–2). When a $\gamma$-ray passes through this slab, three things may occur: (1) the $\gamma$-ray may be completely absorbed by the material; (2) the $\gamma$-ray may be deflected (scattered) with some or no loss of energy; or (3) the $\gamma$-ray may pass through the slab without any interaction. The first two processes together are known as attenuation; the third is known as transmission. The transmission of $\gamma$-rays of a given energy, $E_\gamma$, through a thickness, x, in the above experiment depends only on the nature of the material (density and atomic number) and the thickness, x, of the slab. For a given material, the dependence on the thickness, x, can be experimentally determined by measuring the number of $\gamma$-rays (those without any loss of energy or deflection) transmitted through different thicknesses of the slab.

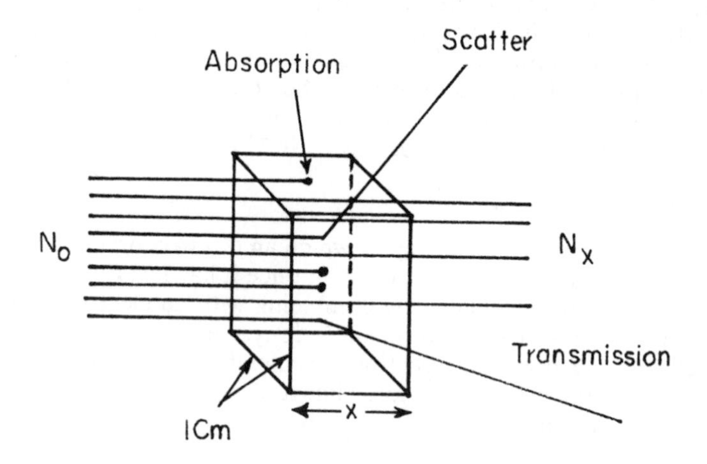


**Fig. 6–2.** Attenuation of $\gamma$-rays. When a flux, $N_0$, of $\gamma$-rays is incident on a substance of x-cm thickness, a smaller flux, $N_x$, is transmitted by the substance. The remaining $\gamma$-rays are either absorbed by the substance or are scattered out from the incident beam.

The data resulting from such an experiment can be accurately related by the following mathematical expression:

$$\frac{N_x}{N_0} = e^{-\mu(\text{linear})x} \qquad [1]$$

where $N_0$ and $N_x$ are, respectively, the number of $\gamma$-rays incident on the slab and transmitted through a thickness, x, of the slab. The linear attenuation coefficient, $\mu$ (linear), represents in physical terms the probability of the interaction of a $\gamma$-ray passing through a unit area of a material 1-cm thick; $\mu$ (linear) strongly depends upon the $\gamma$-ray energy ($E_\gamma$) and the nature of the material (density and atomic number). Note that equation 1 is mathematically the same as that for radioactivity decay with time (see p. 29, Chapter 3). However, here it is describing a completely different physical process—namely, the transmission of $\gamma$-rays through a thickness, x, of a given material. The energy dependence of the linear attenuation coefficient of four substances useful for our purpose is shown in Fig. 6–3.

Experimentally, it is easier to determine the parameter known as the half-value layer (HVL) than to determine $\mu$ (linear). HVL is defined as the thickness of a material which attenuates one-half of the incident $\gamma$-rays. Referring to equation 1, when

$$\frac{N_x}{N_0} = \frac{1}{2},\ x = \text{HVL}$$

HVL is related to $\mu$ (linear) by the following expression:

$$\text{HVL} = \frac{0.693}{\mu(\text{linear})} \qquad [2]$$

which is similar to that for the half-life of a radionuclide ($T_{\frac{1}{2}}$) and its decay constant, $\lambda$. The parameter $\mu$ (linear) is measured in units of $cm^{-1}$, whereas HVL is expressed in cm. Once HVL or $\mu$ (linear) is known, one can easily determine the parallel beam attenuation of $\gamma$-rays through any thickness, x, of a material using equation 1 or one of the shortcuts described in Chapter 3 (p. 33) for solving problems involving exponentials.

*Example:*

The half-value layer of a 140-keV $\gamma$-ray in a NaI(Tl) crystal is approximately 0.3 cm. Determine the percentage of $\gamma$-rays transmitted through a 1.2-cm (≈½″) thick NaI(Tl) crystal for a 140-keV $\gamma$-ray parallel beam.

A thickness of 1.2 cm is expressed as $\frac{1.2}{0.3} = 4$ HVL of a 140-keV $\gamma$-ray in a NaI(Tl) crystal. Since for 1 HVL the ratio of $\frac{N_x}{N_0} = \frac{1}{2}$, for 4 HVL the ratio will be $\frac{N_x}{N_0} = \left(\frac{1}{2}\right)^4 = \frac{1}{16}$; therefore, the percentage of $\gamma$-rays transmitted is $\frac{100}{16}$ or 6%. In other words, 94% of the $\gamma$-rays will be attenuated.

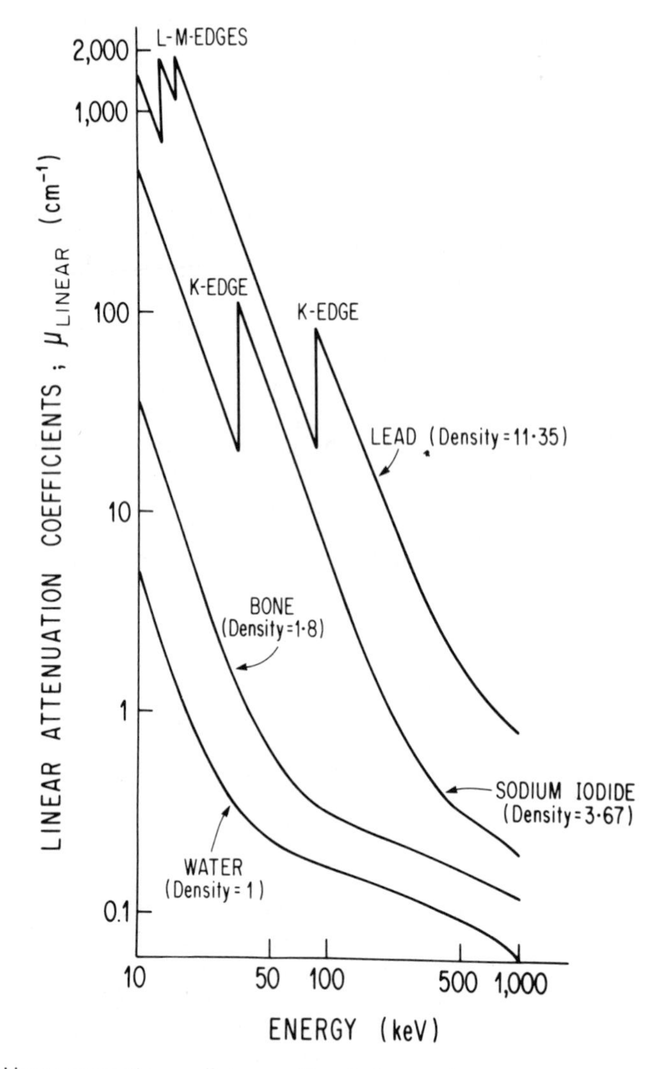


**Fig. 6–3.** Linear attenuation coefficient, $\mu$ (linear) of water, bone, sodium iodide and lead as a function of x- or $\gamma$-ray energy.

Here, the thickness, x, is a multiple of HVL; therefore, it is not necessary to use equation (1).

### Mass Attenuation Coefficient, $\mu$ (mass)

Since $\mu$ (linear) is dependent on the density of the absorbing material, it is often important to know the mass attenuation coefficient, $\mu$ (mass), which removes the effect of the density from the linear attenuation coefficient. The mass attenuation coefficient reflects the probability of an interaction with a unit mass of a material. The mass attenuation coefficient is related to the linear attenuation coefficient by the following expression:

$$\mu \text{ (mass)} = \frac{\mu \text{ (linear)}}{\text{density}}$$

The units for $\mu$ (mass) are expressed as $\frac{\text{cm}^2}{\text{gm}}$.

### Atomic Attenuation Coefficient, $\mu$ (atom)

The two probabilities of interaction, $\mu$ (linear) and $\mu$ (mass), describe the interaction of $\gamma$-rays at the macroscopic level (1 gm or 1 $\text{cm}^3$). How does one relate these two quantities to those at the atomic level (*e.g.*, the probability of $\gamma$-ray interaction with one atom, $\mu$ [atom])? The number of atoms in 1 gm of a substance can be determined by dividing Avogadro's number ($N_{av}$) by the atomic weight (A), *i.e.*, number of atoms in 1 gm $= \frac{N_{av}}{A}$. Then $\mu$ (atom) can be determined by dividing $\mu$ (mass) by the number of atoms in 1 gm of that substance:

$$\mu \text{ (atom)} = \mu \text{ (mass)} \Big/ \frac{N_{av}}{A} = \frac{\mu \text{ (mass)} \cdot A}{N_{av}}$$

The units for $\mu$ (atom) are expressed as $\text{cm}^2$.

### Mechanisms of Interaction

We have described the attenuation of x- or $\gamma$-rays without, thus far, specifying the mechanisms which cause this attenuation. In this energy range (10 keV to 10 MeV), there are three basic processes through which a photon interacts with matter: (1) photoelectric effect, (2) compton effect, and (3) pair production. The relative importance of each type of interaction depends upon the energy of the photon and the atomic number of the material with which it is interacting.

A salient feature of x- or $\gamma$-ray interaction with matter via any of the above three mechanisms is the production of a high-energy charged particle (electron or positron) which then loses energy in the medium by producing ionizations and excitations as previously discussed. Because of this, x- or $\gamma$-radiation is sometimes known as indirectly ionizing radiation.

***Photoelectric Effect:*** When an incident photon interacts with a target atom through the photoelectric effect, it transfers all its energy to one of the electrons in the atom, *i.e.*, the photon is completely absorbed by the atom, as shown in Fig. 6–4. The absorption of energy by the atom leads to its ionization by the emission of an electron, which acquires kinetic energy (Ee) in an amount equal to that of the photon energy (E$\gamma$) minus the electron's binding energy in the shell (B.E.), *i.e.*, Ee = E$\gamma$ − B.E. of the electron. The electron can be released from any atomic shell. If it is released from

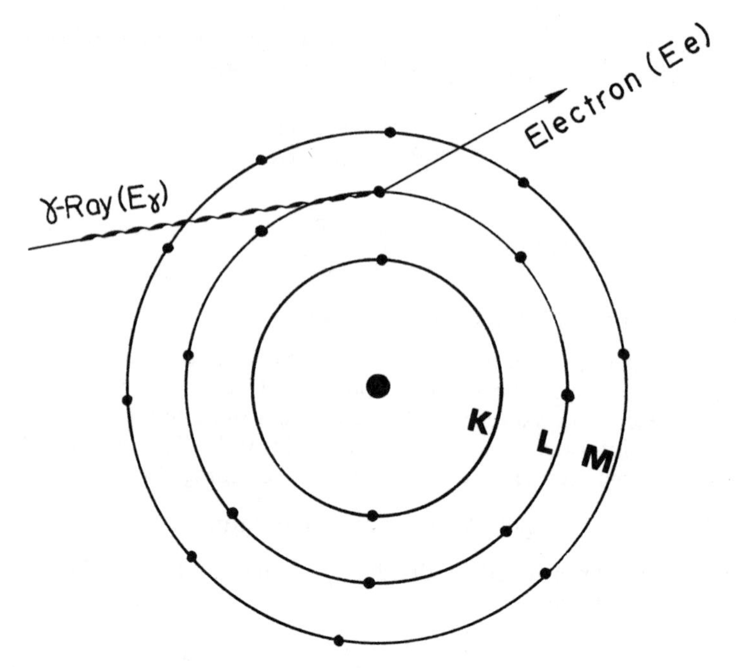


**Fig. 6–4.** Photoelectric interaction. A γ-ray transfers all its energy to an orbital electron of the absorbing material atom and thereby ionizes the atom. The electron thus released carries an amount of energy, Ee, which is equal to the γ-ray energy, $E_\gamma$, minus the binding energy of the electron in that orbit. If during ionization the vacancy is created in an inner orbit, a characteristic x-ray or Auger electron will also follow.

one of the inner shells (*e.g.*, K) a vacancy is created in this shell which is subsequently filled by an electron from one of the higher shells (*e.g.*, L, M), as described in Chapters 1 and 2. This results in the emission of a characteristic x-ray or Auger electron by the atom. If, on the other hand, outer-shell electrons are involved in the photoelectric interaction, the atom is simply ionized. The probability of the photoelectric interaction by an atom, τ (atom), strongly depends on two factors: the energy of the photon Eγ, and the atomic number (Z) of the atom involved in the interaction.

*Dependence on Eγ:* The probability of photoelectric interaction, τ (atom), decreases sharply with the increase in the x- or γ-ray energy. It is inversely proportional to the cube of γ-ray energy, *i.e.*, $\tau\ (\text{atom})\ \alpha\ \frac{1}{E_\gamma{}^3}$. Therefore, the probability of a γ-ray of 45 keV interacting through the photoelectric effect is eight times higher than for a γ-ray of 90-keV energy, *i.e.*, $\frac{(90^3)}{(45^3)} = 8$. There is, however, one exception to this inverse cube law. Whenever the energy of the γ-ray (Eγ) becomes equal to the binding energies of the electrons in the various atomic shells of an atom, the probability of photoelectric interaction rises sharply. For example, the probability of the photoelectric interaction of a 45-keV $\gamma$-ray with a lead atom should be 8 times higher than that of a 90-keV $\gamma$-ray according to the $\frac{1}{E\gamma^3}$ rule. However, since the binding energy of the electrons in the K shell of a lead atom is about 88 keV, the probability of a photoelectric interaction of 90-keV $\gamma$-ray increases to a point where it almost becomes equal to that for a 45-keV $\gamma$-ray. The probability of photoelectric interaction of an 80-keV $\gamma$-ray in lead is about 6 times lower than that of a 90-keV $\gamma$-ray in spite of the fact that the latter is a higher-energy $\gamma$-ray. The regions where the $\frac{1}{E\gamma^3}$ law does not hold are called absorption edges, and their occurrence depends on the atomic number, Z, of the atom with which the $\gamma$-ray interacts. For example, the K absorption edge (where K shell electrons are involved) occurs at approximately 88 keV in lead but at only 32 keV in iodine (Fig. 6–3).

*Dependence on Z:* The photoelectric probability of interaction, $\tau$ (atom), also depends strongly on the atomic number of the atom. It is directly proportional to the fourth power of the atomic number, *i.e.*, $\tau \text{(atom)} \alpha Z^4$. For example, the probability of a $\gamma$-ray of 50 keV interaction with a lead atom ($Z = 82$) is 11,000 times higher than that with an oxygen atom ($Z = 8$), because, $\frac{[82^4]}{[8^4]} = 11{,}000$. Of course, if the $\gamma$-ray energy happens to be near one of the absorption edges of that atom, this rule is modified.

***Compton Effect:*** In this process, a high-energy photon is scattered in billiard-ball fashion by an electron, as shown in Fig. 6–5. The scattered electron gains energy and the incident photon loses energy. The exact amount of the energy gained by the electron or lost by the photon depends on the angle of scatter and the energy of the incident photon $E\gamma$. In general, though, the larger the angle of scatter for a photon, the more energy it loses to an electron. Thus, the maximum transfer of energy from the photon to the electron occurs when the photon is scattered at an angle of 180° (*i.e.*, back-scattered). The energy of the back-scattered photon (minimum energy of scattered photon) is related to the energy of the incident photon ($E\gamma$) by the following expression:

$$E\gamma_{\text{minimum}} \simeq \frac{E\gamma}{1 + 4\,E\gamma} \quad (E\gamma \text{ is in MeV})$$

For a 1-MeV incident photon, the minimum energy of the scattered photon is 0.200 MeV; for a 0.360 MeV-photon, 0.148 MeV; for a 0.140-MeV photon, 0.090 MeV; and for a 0.080-MeV photon, 0.060 MeV. These values of $E_{\text{minimum}}$ are given here as an illustration because of their relevance in $\gamma$-ray energy analysis by NaI (Tl) detectors (Chapter 8). The energies of the scattered photons range from this minimum to that of the incidence photon energy. In Fig. 6–6 we have shown the energy distribution of the scattered photons and scattered electrons for a 360-keV primary photon as a result of compton interaction.

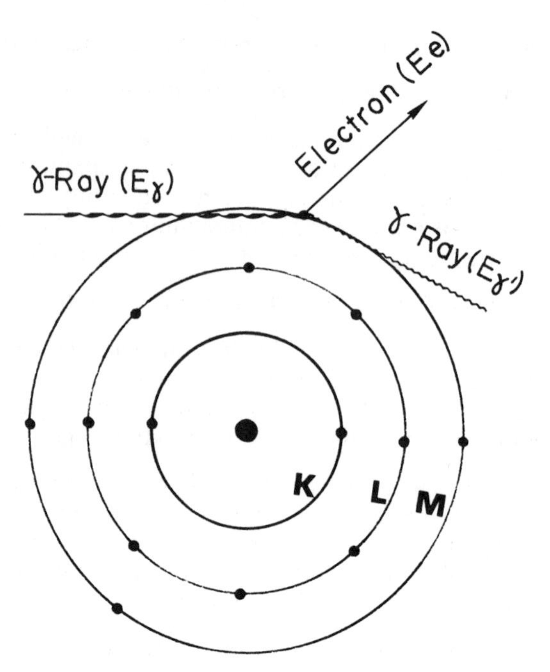


**Fig. 6–5.** Compton interaction. A γ-ray transfers only a partial amount of its energy to an orbital electron (usually the outer shell). The scattered γ-rays carry an energy $E_\gamma'$ which is equal to $E_\gamma$ − Ee, where $E_\gamma$ is the energy of the primary γ-ray and Ee is the energy carried by the scattered electron.

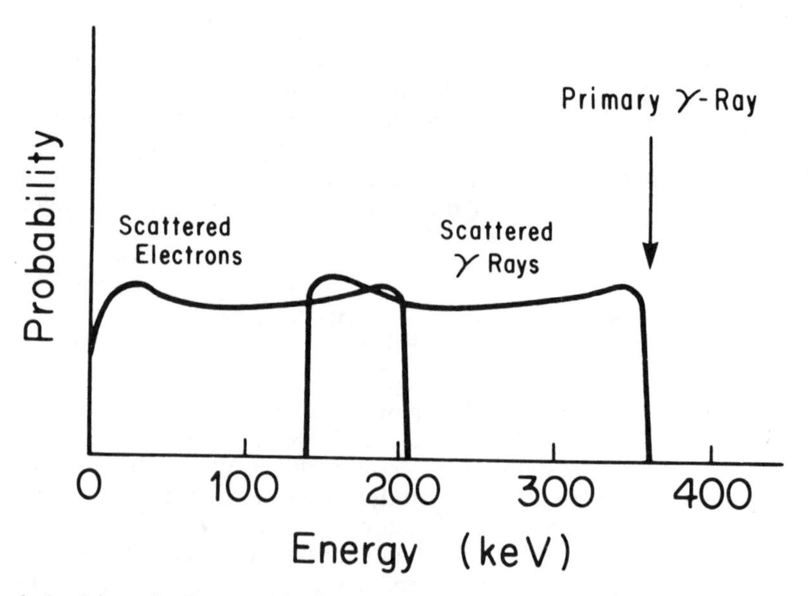


**Fig. 6–6.** Energy distribution of the scattered γ-rays and the scattered electrons in a compton interaction of a 364-keV γ-ray. The maximum amount of energy which a 364-keV γ-ray can transfer to an electron via this process is 210 keV.

*Dependence on E:* The probability of compton interaction, $\sigma$ (atom), initially decreases slowly with the increase in energy and then falls off more rapidly.

*Dependence on Z:* Since each atom contains Z number of electrons, the probability of compton interaction by an atom, $\sigma$ (atom), is directly proportional to the atomic number, *i.e.*, $\sigma$ (atom) $\alpha Z$.

***Pair Production:*** For this interaction to occur, the energy of the $\gamma$-ray must be greater than 1.02 MeV. When a $\gamma$-ray of energy greater than 1.02 MeV passes through the electric field of a nucleus, it creates an electron and a positron (*i.e.*, part of the $\gamma$-ray energy is converted into mass). This process is called pair production and is depicted in Fig. 6–7. The excess energy of the $\gamma$-ray ($E\gamma$ − 1.02 MeV) is shared by $e^-$ and $e^+$ as kinetic energy.

*Dependence on E:* The probability of pair production, $\kappa$ (atom), is zero below 1.02 MeV. At energies higher than 1.02 MeV, $\kappa$ (atom) increases with increases in E and, in fact, becomes the dominant mode of interaction above 10 MeV.

*Dependence on Z:* The probability of pair production, $\kappa$ (atom), for a given atom varies directly as $Z^2$.

## Dependence of $\mu$ (mass) and $\mu$ (linear) on Z

So far we have discussed the dependence of $\mu$ (atom) on Z for different processes of interaction. How do the linear and mass attenuation coefficients vary with atomic number? The linear attenuation of coefficient, $\mu$ (linear), and the mass attenuation coefficient, $\mu$ (mass), depends on Z as $Z^3$, $Z^0$ (*i.e.*, no dependence on Z), and Z for photoelectric, compton and pair production

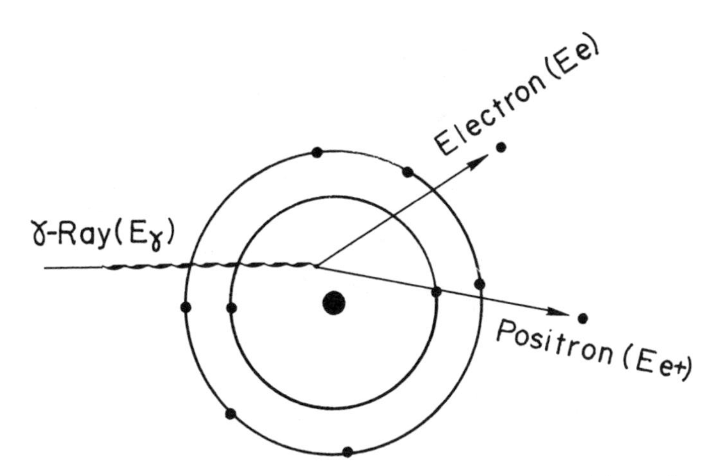


**Fig. 6–7.** Pair production. Under the influence of the positive charge of the nucleus, a $\gamma$-ray whose energy, $E_\gamma$, is more than 1.02 MeV, annihilates and produces a pair of particles (an electron and a positron) which carry the balance of energy, $E_\gamma$ − 1.02 MeV. The positron is subsequently annihilated in the manner described on p. 76.

modes of interaction, respectively. In this respect it is important to remember that when the compton effect is the dominant mode of interaction then each gram of any material (*i.e.*, water, iodine, bone, or lead) attenuates the $\gamma$- or x-rays to about the same extent. However, attenuation per $cm^3$ of those materials will still differ, in proportion to their density.

### Relative Importance of the Three Processes

The total interaction probability of an atom, $\mu$ (atom), then, is the sum of the three probabilities, $\tau$ (atom), $\sigma$ (atom) and $\kappa$ (atom), *i.e.*, $\mu$ (atom) = $\tau$ (atom) + $\sigma$ $(\text{atom})_\gamma$ + $\kappa$ (atom). Because of the complex and varying dependence of $\mu$ (atom) on $\gamma$-ray energy $E\gamma$ and the atomic number Z of the material, generally one of these processes becomes the dominant mode of interaction for a given $\gamma$-ray energy and atomic number of the material. Fig. 6–8 shows the relative importance of these processes for the 10-keV to 1-MeV energy range for four materials of great importance in nuclear medicine: water (tissue type), bone, sodium iodide [NaI(Tl)], and lead (Pb). In water

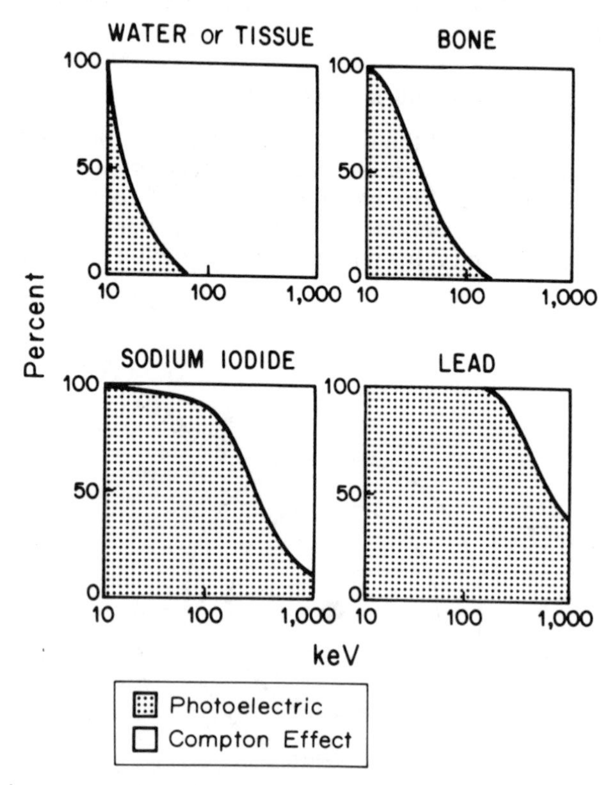


**Fig. 6–8.** Relative contributions of photoelectric and compton interaction to the total attenuation coefficient of water (tissue), bone, sodium iodide and lead as a function of energy. In water and bone, almost all interactions above 50 keV are via the compton effect, whereas in sodium iodide and lead the photoelectric effect is the dominant mode of interaction even up to 300 keV.

or bone, most of the interaction of x- or $\gamma$-rays above 50 keV is through the compton effect. For lead, photoelectric interaction is the dominant mode of interaction even up to 1 MeV, although at these energies compton interaction does become important. In NaI(Tl), photoelectric effect is the dominant mode of interaction up to 300 keV.

## INTERACTION OF NEUTRONS

Since neutrons do not have any electric charge, they do not experience the attractive or repulsive forces experienced by charged particles. Instead, these interact in billiard-ball fashion (direct hit) with the nuclei. When the energy of the neutrons is very low, they can easily enter a nucleus and may form radionuclides. Interaction of neutrons does not play a significant role in nuclear medicine, except in the production of some radionuclides by neutron-capture reactions (Chapter 4), and therefore is not discussed here.

## PROBLEMS

1. Arrange the following radiations in order of their penetrability: 1 MeV x-ray, 1 MeV electron, 1 MeV $\alpha$ particle, 50 keV characteristic x-ray, 400 keV $\gamma$-ray.
2. List the factors on which the range of a charged particle depends.
3. If you know the range of a charged particle in water, what other information do you need to determine its range in another medium? If you know the linear attenuation coefficient of a gamma ray in water, will the same information let you determine the linear attenuation coefficient of another medium for the same $\gamma$-ray?
4. The linear attenuation coefficients of 300 keV $\gamma$-rays for tissue, sodium iodide, and lead are approximately 0.12, 0.8, and 4.5 $cm^{-1}$ respectively. Determine the half-value layer for each material.
5. Photoelectric effect generally decreases sharply with $\gamma$-ray energy. In what areas (energies) does this law break down?
6. Determine the energy of a backscattered $\gamma$-ray if the energy of the primary $\gamma$-ray is one of the following: 100 keV, 500 keV, or 1 MeV.

# 7

# Radiation Dosimetry

Radiation is known to produce a number of deleterious effects in a living system. Therefore, it is important to properly assess the benefits and risks to a patient from a given nuclear-medicine procedure, insuring that the benefits outweigh the risks, if any, by a significant margin. Since one of the factors which strongly influences the intensity or probability of radiation effects is the radiation dose, this assessment can be performed only if one knows the radiation dose that will be delivered to a patient by a particular procedure. This chapter describes the various methods for determining the radiation dose to a patient from internally administered radionuclides or radiopharmaceuticals.

In nuclear-medicine procedures, it is almost impossible to measure the radiation dose directly using any kind of radiation detector. Instead, this has to be calculated by using a variety of physical and biological data and mathematical equations specially developed for this purpose. It should be emphasized, however, at the outset, that the accuracy of these calculations is complicated by several factors. (1) It is difficult to obtain the biological data needed for these calculations with sufficient accuracy as these are, in practice, extrapolated from animal data or in some cases from very limited human data. (2) A variety of assumptions routinely made in these calculations (for example, uniform distribution, instant uptake of the radiopharmaceutical in the given organ, or single exponential biological elimination) seldom holds true in practice. (3) Radiation-dose calculations are usually addressed to a hypothetical man known as the "standard man." Various particulars such as the total body weight and weight of various organs of the standard man are given in Table 7–1. These may differ appreciably in an actual case.

All these factors, when combined, make radiation-dose calculations susceptible to large errors. Hence, the radiation-dose calculations presented here and elsewhere most likely represent an average dose from a given procedure which may vary by a factor of two, or even more, in an individual case.

## DEFINITIONS

Obviously, to calculate the radiation dose one should know the meaning of radiation dose and radiation-dose rate.

**Table 7–1. Organ Masses of a Standard Man**

| Organ | Mass (gm) |
|---|---|
| Total Body | 70,000 |
| Bladder | 509 |
| Kidneys (both) | 288 |
| Liver | 1,833 |
| Lungs | 999 |
| Ovaries | 8.8 |
| Pancreas | 61 |
| Skeleton with Marrow | 10,091 |
| Spleen | 176 |
| Stomach | 402 |
| Testicles | 38 |
| Thyroid | 20 |

***Radiation Dose, D:*** Radiation dose, more precisely called radiation absorbed dose, is a measure of the total energy absorbed from the radiation by one gram of a substance. Its unit is the rad, which is a short notation for the radiation absorbed dose. A rad is defined as 100 ergs of absorbed energy per gm of a tissue or substance. If one knows the amount of energy absorbed in ergs/gm, then to obtain the radiation dose one divides it by 100 (*i.e.*, 1 rad = 100 ergs/gm).

In the SI Unit System, rad is replaced by gray (Gy). A gray equals 100 rad or a rad equals $10^{-2}$ gray.

***Radiation Dose Rate,*** $\frac{dD}{dt}$***:*** Radiation dose rate, $\frac{dD}{dt}$, is defined as the amount of energy absorbed per unit time per gram of tissue. Its units may be expressed in various ways, such as rads (Gy) per minute, rads (Gy) per hour, or rads (Gy) per day.

## PARAMETERS OR DATA NEEDED

In a typical situation in nuclear medicine, a known amount of radioactivity of a radiopharmaceutical is administered to a patient. A certain fraction (f) of the radiopharmaceutical is then localized in the organ of interest. One is interested in knowing the radiation dose delivered to this organ and sometimes to various other organs, as well. Two types of data are required for these calculations, one related to the decay characteristics of the radionuclide and the other to the biological distribution and elimination of the radiopharmaceutical. Table 7–2 lists the various parameters needed for these calculations, together with the units (I have retained the old units here because a large amount of the data needed is still in old units.) and symbols employed here. Some of the parameters have been defined in Chapters 2 and 3; some are self-evident, and the others are discussed in this chapter.

**Table 7–2. Various Parameters and Symbols Used in the Calculation of Radiation Dose**

| *Parameter* | *Symbol* | *Units* |
|---|---|---|
| Any Given Radiation | $i$ | — |
| Energy of Radiation i | $E_i$ | MeV |
| Frequency of Emission of Radiation i | $n_i$ | Per decay |
| Equilibrium Dose Constant for Radiation i | $\Delta_i$ | g.rad/μCi.h |
| Total Number of Radiations | $n$ | — |
| Absorbed Fraction | $\phi_i(T \leftarrow S)$ | — |
| Self-Absorbed Fraction | $\phi_i$ | — |
| Target Organ | $T$ | — |
| Source Organ | $S$ | — |
| Dose Rate at Time t | $\frac{dD}{dt}$ | rad/hr |
| Dose (Absorbed) | $D$ | rad |
| Radioactivity at Time t | $A(t)$ | μCi |
| Radioactivity at Time 0 | $A_0$ | μCi |
| Mass of Target Organ | $M$ | gm |
| Physical Half-Life | $T_{\frac{1}{2}}$ | hr |
| Biological Half-Life | $T_{\frac{1}{2}}(Bio)$ | hr |
| Effective Half-Life | $T_{\frac{1}{2}}(eff)$ | hr |
| Fraction of Radioactivity Localized in an Organ | $f$ | — |

## CALCULATION OF THE RADIATION DOSE

In order to calculate the radiation dose one has to determine the average amount of energy absorbed by one gram of tissue of a target (organ of interest) from the total energy released by the decay of a given amount of radioactivity. The radioactivity may be distributed within the target volume (T) or in a source volume (S), which is outside the target volume (T). This is accomplished through the following four steps:

1. Calculation of the rate of energy emission (ergs/hr) by the radionuclidic distribution in the form of various types of radiation.
2. Calculation of the rate of energy absorption by the target volume.
3. Calculation of the average Dose Rate $\frac{dD}{dt}$.
4. Calculation of the average Dose D.

This method of radiation dose calculation is known as the absorbed fraction method. The first three steps require mainly physical data such as decay characteristics, organ shape and size, etc., whereas the fourth step requires biological distribution data.

### Rate of Energy Emission

Let us first consider a radionuclide which emits only one type of radiation (fractional emission frequency = 1) of energy E (MeV) per decay. One microcurie ($3.7 \times 10^4$ decay/sec) of this radionuclide will, therefore, emit energy at a rate of $3.7 \times 10^4 \times E$ MeV/sec/μCi. If we change the units of energy from MeV to ergs (1 MeV = $1.6 \times 10^{-6}$ erg) and the units of time from seconds to hours (1 hr = 3600 sec), the rate of energy emission by one microcurie of this radionuclide becomes equal to $3.7 \times 10^4 \times 1.6 \times 10^{-6} \times 3600 \times E$ ergs/hr/$\mu$Ci, or 213 E ergs/hr/$\mu$Ci.

In the case of a radionuclide which emits more than one radiation, say 1, 2, 3 . . . n, with fractional emission frequencies $n_1$, $n_2$, $n_3$ . . . $n_n$ and energies of $E_1$, $E_2$, $E_3$ . . . $E_n$, respectively, the rate of energy emission for each type of radiation from 1 $\mu$Ci of activity will be equal to 213 $n_1$ $E_1$ ergs/hr/$\mu$Ci for radiation 1, 213 $n_2$ $E_2$ ergs/hr/$\mu$Ci for radiation 2, and so on.

### Rate of Energy Absorption

To calculate the rate of absorption of energy by a target volume (T) from a radionuclidic distribution in a source volume (S) we have to define a new quantity known as the absorbed fraction, $\phi_i(T \leftarrow S)$. The absorbed fraction $\phi_i(T \leftarrow S)$ is defined as the ratio of energy absorbed by a target volume (T) from a radiation i to the amount of energy released by a radionuclidic distribution in volume (S) in the form of radiation i. In other words,

$$\phi_i(T \leftarrow S) = \frac{\text{Amount of energy absorbed in volume T from radiation i}}{\text{Amount of energy emitted in volume S as radiation i}}$$

In the majority of problems encountered in nuclear medicine, the radioactivity is distributed within the target volume T itself (*i.e.*, T is the same as S). In these cases, the absorbed fraction is known as the self-absorbed fraction and is expressed simply as $\phi_i$, meaning that the target volume and source volume are the same.

Once $\phi_i(T \leftarrow S)$ is known, the rate of energy absorption by a target volume T is simply obtained by multiplying the rate of energy emission of radiation i (Step 1) by the absorbed fraction $\phi_i(T \leftarrow S)$:

Rate of energy absorption by target volume T from radiation i equals Rate of energy emission in the form of radiation i in the source volume, S, times $\phi_i(T \leftarrow S)$

or

$$= 213\ n_i E_i \phi_i(T \leftarrow S) \text{ ergs/hr/}\mu\text{Ci}$$

If there are n radiations, the rate of the total energy absorption will be equal to the sum of the energies absorbed from each radiation, *i.e.*, $= 213\ n_1E_1\ \phi_1(T \leftarrow S) + 213\ n_2E_2\phi_2(T \leftarrow S) + \ldots .\ 213\ n_nE_n\phi_n(T \leftarrow S)$ ergs/hr/$\mu$Ci. The above expression can be written in a concise form as

$$213 \sum_{i=1}^{n} n_iE_i\phi_i(T \leftarrow S) \text{ ergs/hr/}\mu\text{Ci},$$

where

$\sum_{i=1}^{n}$ is the summation of all the terms when i is changed from 1 to n.

**Table 7–3. Absorbed Fraction $\phi_i$ for Different $\gamma$-ray Energies and Various Organs**

| | Energy (keV) | | | | | | |
|---|---|---|---|---|---|---|---|
| *Organ* | *15* | *30* | *50* | *100* | *200* | *500* | *1000* |
| Bladder | 0.885 | 0.464 | 0.201 | 0.117 | 0.116 | 0.116 | 0.107 |
| Stomach | 0.860 | 0.414 | 0.176 | 0.101 | 0.101 | 0.101 | 0.093 |
| Kidneys | 0.787 | 0.298 | 0.112 | 0.066 | 0.068 | 0.073 | 0.067 |
| Liver | 0.898 | 0.543 | 0.278 | 0.165 | 0.158 | 0.157 | 0.144 |
| Lungs | 0.665 | 0.231 | 0.089 | 0.049 | 0.050 | 0.051 | 0.045 |
| Pancreas | 0.666 | 0.195 | 0.068 | 0.038 | 0.042 | 0.044 | 0.040 |
| Skeleton | 0.893 | 0.681 | 0.400 | 0.173 | 0.123 | 0.118 | 0.110 |
| Spleen | 0.817 | 0.331 | 0.128 | 0.071 | 0.073 | 0.077 | 0.070 |
| Thyroid | 0.592 | 0.149 | 0.048 | 0.028 | 0.031 | 0.032 | 0.029 |
| Total Body | 0.933 | 0.774 | 0.548 | 0.370 | 0.338 | 0.340 | 0.321 |

How does one determine $\phi_i(T \leftarrow S)$? Determination of the absorbed fraction requires the exact knowledge of the interaction of radiation with matter which was discussed briefly in the previous chapter. In the case of particulate radiations such as $\beta$ particles, conversion electrons, or $\alpha$ particles, almost all of the energy emitted by a radionuclide is absorbed in the volume of distribution itself, provided the source volume is larger than 1 cm$^3$. Then, $\phi_i(T \leftarrow S) = 0$, unless T and S are the same, in which case $\phi_i = 1$. This also holds true for x- or $\gamma$-radiation with energies less than 10 keV.

For $\gamma$- or x-radiations with energies higher than 10 keV, the absorbed fraction $\phi_i(T \leftarrow S)$ strongly depends on: (1) the energy of the x- or $\gamma$-ray; (2) the shape and size of the source volume; and (3) the shape, size, and distance of the target volume. In general, $\phi_i(T \leftarrow S)$ is always less than or equal to 1. Its value first decreases with an increase in the energy of the x- or $\gamma$-ray and then eventually levels off.

Exact computation of $\phi_i(T \leftarrow S)$ from the basic mechanisms of interaction of x- or $\gamma$-rays with matter requires the use of large computers. The *Journal of Nuclear Medicine** has published a variety of tables listing $\phi_i(T \leftarrow S)$ for different x- or $\gamma$-ray energies, and source and target volumes. Table 7–3 lists the absorbed fraction for various organs of a standard man for different x- or $\gamma$-ray energies, when the radionuclidic distribution is within the same organ (*i.e.*, T is the same as S). For other combinations the reader is referred to the original articles.*

## Dose Rate, $\frac{dD}{dt}$

If one now divides the rate of energy absorption by the target by its mass (M), this will give the rate of energy absorption per gram of tissue which when divided by 100 (to convert erg/gm to rad) yields the dose rate for each

* Suppl. No. 1, 1968; Suppl. No. 3, 1969; Suppl. No. 5, 1971.

microcurie of activity: that is, the dose rate per $\mu$Ci of activity will be equal to

$$\frac{213 \sum_{i=1}^{n} n_i E_i \phi_i(T \leftarrow S)}{100 \times M} \text{ rad/hr/}\mu\text{Ci}$$

or

$$= \frac{2.13}{M} \sum_{i=1}^{n} n_i E_i \phi_i(T \leftarrow S) \text{rad/hr/}\mu\text{Ci}$$

If the source volume contains A(t) $\mu$Ci at that time (t), then the dose rate $\frac{dD}{dt}$ from A(t) amount of radioactivity becomes

$$\frac{dD}{dt} = \frac{2.13}{M} \cdot A(t) \cdot \sum_{i=1}^{n} n_i E_i \phi_i(T \leftarrow S) \text{ rad/hr} \quad [1]$$

**Average Dose, D**

The radioactivity A(t) in a biological system is being continuously eliminated, with an effective half-life of $T_{\frac{1}{2}}(\text{eff})$; *i.e.*,

$$A(t) = A_0 \exp\left(-\frac{0.693t}{T_{\frac{1}{2}}(\text{eff})}\right) \mu\text{Ci} \quad [2]$$

Therefore, the dose rate $\frac{dD}{dt}$ is continuously decreasing with time and eventually becomes zero. How does one compute the total dose to the patient from the time of administration (t = 0) to the time when the dose rate has finally been reduced to zero? For this one has to integrate the dose rate $\frac{dD}{dt}$ from 0 time to infinite time, or $D = \int_0^\infty \frac{dD}{dt} \cdot dt$. When this integration is performed, the radiation dose is given by the following expression:

$$D(T \leftarrow S) = 2.13 \frac{A_0}{M} \cdot [1.44\ T_{\frac{1}{2}}(\text{eff})] \cdot \sum_{i=1}^{n} n_i E_i \phi_i(T \leftarrow S) \text{ rad} \quad [3]$$

By defining $\Delta_i = 2.13\ n_i E_i$ (rad g/$\mu$Ci·h) [4]

Equation (3) can be further reduced to

$$D(T \leftarrow S) = \frac{A_0}{M} \cdot [1.44 \cdot T_{\frac{1}{2}}\ (\text{eff})] \cdot \sum_{i=1}^{n} \Delta_i \phi_i(T \leftarrow S) \text{ rad} \quad [5]$$

In the case where the target and source volume are the same, the self dose D is given by

$$D = \frac{A_0}{M}\cdot[1.44\cdot T_{\frac{1}{2}}(\text{eff})]\cdot\sum_{i=1}^{n} \Delta_i \phi_i \text{ rad} \qquad [6]$$

From this expression it is evident that to minimize the radiation dose to a patient one has to use either a smaller amount of radioactivity ($A_0$) or a radiopharmaceutical with a short $T_{\frac{1}{2}}$(eff) [$T_{\frac{1}{2}}$(eff) will be short if either $T_{\frac{1}{2}}$ or $T_{\frac{1}{2}}$(Bio) is short].

In general, when one calculates the radiation dose to an organ, one has to consider the self dose from the radioactive distribution in the organ itself as well as the contribution to the radiation dose from the radioactive distribution in the other organs, if any. In the latter case, only the x- or $\gamma$-rays will contribute to the radiation dose.

Also, I should point out that the above equations 5 and 6 assume that the uptake in the organ is instantaneous and the disappearance of the radioactivity from the source can be expressed by a single exponential term. This does not have to be so. Under that circumstance, the exact time activity curve should be used to calculate the cumulated radioactivity $\tilde{A}$, which is defined as $\tilde{A} = \int_0^\infty A(t)dt$. Since these calculations are more complicated, we shall not go into the details here.

## Simplification of Radiation Dose Calculations Using 'S' Factors

Recently, in an attempt to simplify the calculation of radiation dose in routine clinical situations, Snyder and his colleagues combined the physical data such as radiations emitted by a radionuclide and the absorbed fractions for various source and target combinations of a standard man into one single term which they called 'S' factor. S is defined as follows:

$$S(T \leftarrow S) = \frac{2.13}{M}\cdot\sum_{i=1}^{n} n_i E_i \phi_i \text{ rad}/\mu\text{Ci}\cdot\text{hr}$$

Substitution of S factor in equations 1 and 5 simplifies them to the following equations respectively.

$$\frac{dD}{dT} = A(t)\cdot S(T \leftarrow S) \text{ rad/hr} \qquad [7]$$

and

$$D(T \leftarrow S) = \tilde{A}\cdot S(T \leftarrow S) \quad \text{or,} \qquad [8]$$
$$= 1.44\, A_0\cdot T_{\frac{1}{2}}(\text{eff})\cdot S(T \leftarrow S) \text{ rad}$$

S factor depends upon only the physical data and is unique for a given radionuclide and a pair of organs of a standard man. Hence, it has to be calculated only once for each radionuclide and each pair of organs (source and target). Snyder and his colleagues have calculated these factors for a large number of medically useful radionuclides. These are published as Supplement 11, *Journal of Nuclear Medicine*. Table 7–4 lists some of the S factors for $^{99m}$Tc for a number of organ pairs of a standard man. For other combinations of pairs and radionuclides, the original reference should be consulted (available only in old units).

Knowing S factors, the calculation of the radiation dose is simple. One plugs the values of the radioactivity localized in the source organ, the effective half-life, and the S factor in equation 8 and one obtains the radiation dose.

*Examples:*

***Problem 1:*** A liver scan is performed on a patient using 2 mCi of $^{99m}$Tc-labeled sulphur colloid. Assuming that 90% of the injected dose is localized in the liver instantaneously and that the effective half-life $T_{\frac{1}{2}}$(eff) is equal to 6 hr, calculate the radiation dose delivered to the liver of the patient.

Calculation: Since mass of the liver is not given in the problem, we shall assume it to be that of a standard man. Therefore, M = 1800 gm. Initial amount of activity localized in the liver, $A_0$

$$= 0.90 \times 2000\ \mu\text{Ci}$$

$$= 1800\ \mu\text{Ci (66.6 MBq)}$$

$$\text{and} \qquad T_{\frac{1}{2}}(\text{eff}) = 6\ \text{hr}$$

Parameters such as $n_i$, $E_i$, $\Delta_i$ for $^{99m}$Tc, and $\phi_i$ for liver for $^{99m}$Tc radiations and $\Sigma \Delta_i \phi_i$ are given in Table 7–5.

Substituting these values in equation 6 we get

$$\text{Dose (liver)} = \frac{1800}{1800} \times 1.44 \times 6 \times 0.078$$

$$= 0.67\ \text{rad (6.7 mGy)}$$

The remaining 10% of the dose, if not excreted in a short time from the body, will also contribute some radiation dose to the liver. Its contribution, however, will be very small and, therefore, is ignored here.

***Problem 2:*** Calculate the radiation dose to the liver, bone marrow (red), ovaries and testes using S factors and the data given in problem 1.

As in problem 1,

$$T_{\frac{1}{2}}(\text{eff}) = 6\ \text{hr},\ A_0 \text{ in liver} = 1800\ \mu\text{Ci (66.6 MBq)}$$

From Table 7–4

$$S\ (\text{liver} \leftarrow \text{liver}) = 4.6 \times 10^{-5}$$

$$S\ (\text{marrow} \leftarrow \text{liver}) = 1.6 \times 10^{-6}$$

$$S\ (\text{ovaries} \leftarrow \text{liver}) = 4.5 \times 10^{-7}$$

$$S\ (\text{testes} \leftarrow \text{liver}) = 6.2 \times 10^{-8}$$

**Table 7–4. 'S' Factors for $^{99m}Tc$ for Various Combinations of Target and Source Organs of a Standard Man**

| *Target* / *Source* | *Bladder Content* | *Stomach Content* | *Kidneys* | *Liver* | *Lung* | *Marrow Red* | *Bone (Av)* | *Spleen* | *Thyroid* | *Total Body* |
|---|---|---|---|---|---|---|---|---|---|---|
| Bladder Wall | $1.6 \times 10^{-4}$ | $2.7 \times 10^{-7}$ | $2.8 \times 10^{-7}$ | $1.6 \times 10^{-7}$ | $3.6 \times 10^{-8}$ | $9.9 \times 10^{-7}$ | $5.1 \times 10^{-7}$ | $1.2 \times 10^{-7}$ | $2.1 \times 10^{-9}$ | $2.3 \times 10^{-6}$ |
| Bone (Total) | $9.2 \times 10^{-7}$ | $9.0 \times 10^{-7}$ | $1.4 \times 10^{-6}$ | $1.1 \times 10^{-6}$ | $1.5 \times 10^{-6}$ | $4.0 \times 10^{-6}$ | $1.1 \times 10^{-5}$ | $1.1 \times 10^{-6}$ | $1.0 \times 10^{-6}$ | $2.5 \times 10^{-6}$ |
| Stomach Wall | $2.7 \times 10^{-7}$ | $1.3 \times 10^{-4}$ | $3.6 \times 10^{-6}$ | $1.9 \times 10^{-6}$ | $1.8 \times 10^{-6}$ | $9.5 \times 10^{-7}$ | $5.5 \times 10^{-7}$ | $1.0 \times 10^{-5}$ | $4.5 \times 10^{-8}$ | $2.2 \times 10^{-6}$ |
| Kidneys | $2.6 \times 10^{-7}$ | $3.5 \times 10^{-6}$ | $1.9 \times 10^{-4}$ | $3.9 \times 10^{-6}$ | $8.4 \times 10^{-7}$ | $2.2 \times 10^{-7}$ | $8.2 \times 10^{-7}$ | $9.1 \times 10^{-6}$ | $3.4 \times 10^{-8}$ | $2.2 \times 10^{-6}$ |
| Liver | $1.7 \times 10^{-7}$ | $2.0 \times 10^{-6}$ | $3.9 \times 10^{-6}$ | $4.6 \times 10^{-5}$ | $2.5 \times 10^{-6}$ | $9.2 \times 10^{-7}$ | $6.6 \times 10^{-7}$ | $9.8 \times 10^{-7}$ | $9.3 \times 10^{-8}$ | $2.2 \times 10^{-6}$ |
| Lungs | $2.4 \times 10^{-8}$ | $1.7 \times 10^{-6}$ | $8.5 \times 10^{-7}$ | $2.5 \times 10^{-6}$ | $5.2 \times 10^{-5}$ | $1.2 \times 10^{-6}$ | $9.4 \times 10^{-7}$ | $2.3 \times 10^{-6}$ | $9.4 \times 10^{-7}$ | $2.0 \times 10^{-6}$ |
| Marrows | $2.2 \times 10^{-6}$ | $1.6 \times 10^{-6}$ | $3.8 \times 10^{-6}$ | $1.6 \times 10^{-6}$ | $1.9 \times 10^{-6}$ | $3.1 \times 10^{-5}$ | $6.6 \times 10^{-6}$ | $1.7 \times 10^{-6}$ | $1.1 \times 10^{-6}$ | $1.1 \times 10^{-6}$ |
| Ovaries | $7.3 \times 10^{-6}$ | $5.0 \times 10^{-7}$ | $1.1 \times 10^{-6}$ | $4.5 \times 10^{-7}$ | $9.4 \times 10^{-8}$ | $3.2 \times 10^{-6}$ | $8.5 \times 10^{-7}$ | $4.0 \times 10^{-7}$ | $4.9 \times 10^{-9}$ | $2.4 \times 10^{-6}$ |
| Spleen | $6.6 \times 10^{-7}$ | $1.8 \times 10^{-5}$ | $8.6 \times 10^{-6}$ | $9.2 \times 10^{-7}$ | $2.3 \times 10^{-6}$ | $9.2 \times 10^{-7}$ | $5.8 \times 10^{-7}$ | $3.3 \times 10^{-4}$ | $1.1 \times 10^{-7}$ | $2.2 \times 10^{-6}$ |
| Testes | $4.7 \times 10^{-6}$ | $5.1 \times 10^{-8}$ | $8.8 \times 10^{-8}$ | $6.2 \times 10^{-8}$ | $7.9 \times 10^{-9}$ | $4.5 \times 10^{-7}$ | $6.4 \times 10^{-7}$ | $4.8 \times 10^{-8}$ | $5.0 \times 10^{-6}$ | $1.7 \times 10^{-6}$ |
| Thyroid | $2.1 \times 10^{-9}$ | $8.7 \times 10^{-8}$ | $4.8 \times 10^{-8}$ | $1.5 \times 10^{-7}$ | $9.2 \times 10^{-7}$ | $6.8 \times 10^{-7}$ | $7.9 \times 10^{-7}$ | $8.7 \times 10^{-8}$ | $2.3 \times 10^{-3}$ | $1.5 \times 10^{-6}$ |
| Total Body | $1.9 \times 10^{-6}$ | $1.9 \times 10^{-6}$ | $2.2 \times 10^{-6}$ | $2.2 \times 10^{-6}$ | $2.0 \times 10^{-6}$ | $2.2 \times 10^{-6}$ | $6.6 \times 10^{-7}$ | $2.2 \times 10^{-6}$ | $1.8 \times 10^{-6}$ | $2.0 \times 10^{-6}$ |

**Table 7–5. Calculation of $\sum \Delta_i \phi_i$: Problem 1 ($^{99m}Tc$)**

| *Radiation i* | $n_i$ | $E_i$ | $\Delta_i^*$ | $\phi_i$ | $\Delta_i \phi_i$ |
|---|---|---|---|---|---|
| Gamma 1 | | | | | |
| (Conversion electrons only) | 0.986 | 0.002 | 0.004 | 1 | 0.004 |
| Gamma 2 | 0.883 | 0.140 | 0.264 | 0.16 | 0.042 |
| K Conversion Electron | 0.088 | 0.119 | 0.022 | 1 | 0.022 |
| L Conversion Electron | 0.011 | 0.138 | 0.003 | 1 | 0.003 |
| M Conversion Electron | 0.004 | 0.140 | 0.001 | 1 | 0.001 |
| Gamma 3 | | | | | |
| (Conversion electrons only) | 0.01 | 0.122 | 0.003 | 1 | 0.003 |
| K (α) X-Ray | 0.064 | 0.018 | 0.003 | 0.88 | 0.0026 |
| K (β) X-Ray | 0.012 | 0.021 | — | 0.87 | — |
| KLL Auger Electron | 0.015 | 0.015 | — | 1 | — |
| LMM Auger Electron | 0.106 | 0.002 | — | 1 | — |
| MXY Auger Electron | 1.23 | 0.0004 | — | 1 | — |
| | | | | $\sum_i \Delta_i \phi_i$ = | 0.078† |

* $\Delta_i$ calculated using eq. 4.
† Sum of all $\Delta_i \phi_i$.

Substituting these values in equation 8.

$$D\ (\text{liver} \leftarrow \text{liver}) = 1.44 \times 1800 \times 6 \times 4.6 \times 10^{-5} = 0.72\ \text{rad}$$

$$D\ (\text{marrow} \leftarrow \text{liver}) = 1.44 \times 1800 \times 6 \times 1.6 \times 10^{-6} = 0.025\ \text{rad}$$

$$D\ (\text{ovaries} \leftarrow \text{liver}) = 1.44 \times 1800 \times 6 \times 4.5 \times 10^{-7} = 0.007\ \text{rad}$$

$$D\ (\text{testes} \leftarrow \text{liver}) = 1.44 \times 1800 \times 6 \times 6.2 \times 10^{-8} = 0.001\ \text{rad}\ (0.01\ \text{mGy})$$

This problem clearly illustrates the simplicity of 'S' factor approach in the calculation of the radiation dose. Again we have neglected the contribution to the radiation dose of the 10% radiopharmaceutical dose not localized in the liver. Also note that the dose to the liver by this method is slightly higher than that in problem 1. This in part is due to the fact that 'standard' man data used for the calculation of 'S' factors is slightly different than that used to compute the original absorbed fraction ($\phi_i$).

Caution: S factor method cannot be applied when the radiation dose is calculated for persons who appreciably differ from a 'standard' man such as children.

***Problem 3:*** A patient is treated for hyperthyroidism with a 5-mCi dose of

**Table 7–6. Calculation of $\sum \Delta_1 \phi_1$: Problem 3 ($^{131}I$)**

| *Radiation i* | $n_i$ | $E_i$ | $\Delta_i$ | $\phi_i$ | $\Delta_i \phi_i$ |
|---|---|---|---|---|---|
| Beta 1 | 0.016 | 0.070 | 0.002 | 1 | 0.002 |
| Beta 2 | 0.069 | 0.095 | 0.014 | 1 | 0.014 |
| Beta 3 | 0.005 | 0.143 | 0.001 | 1 | 0.001 |
| Beta 4 | 0.905 | 0.192 | 0.369 | 1 | 0.369 |
| Beta 5 | 0.006 | 0.286 | 0.004 | 1 | 0.004 |
| Gamma 1 | 0.017 | 0.080 | 0.003 | 0.035 | — |
| K Conversion Electron | 0.029 | 0.046 | 0.003 | 1 | 0.003 |
| Gamma 2 | | | | | |
| (Conversion Electron) | 0.004 | 0.129 | 0.001 | 1 | 0.001 |
| Gamma 3 | 0.047 | 0.284 | 0.029 | 0.03 | 0.001 |
| K Conversion Electron | 0.002 | 0.250 | 0.001 | 1 | 0.001 |
| Gamma 4 | 0.002 | 0.326 | 0.001 | 0.03 | — |
| Gamma 5 | 0.833 | 0.364 | 0.646 | 0.03 | 0.019 |
| K Conversion Electron | 0.017 | 0.330 | 0.012 | 1 | 0.012 |
| L Conversion Electron | 0.003 | 0.359 | 0.002 | 1 | 0.002 |
| Gamma 6 | 0.003 | 0.503 | 0.003 | 0.03 | — |
| Gamma 7 | 0.069 | 0.637 | 0.093 | 0.03 | 0.003 |
| Gamma 8 | 0.016 | 0.723 | 0.025 | 0.03 | 0.001 |
| K ($\alpha$) X-Rays | 0.038 | 0.030 | 0.002 | 0.15 | — |
| | | | | $\sum_i \Delta_i \phi_i$ = | 0.433* |

* Sum of all $\Delta_i \phi_i$.

$^{131}I$. Calculate the radiation dose that will be delivered to the thyroid of the patient, assuming that the estimated weight of the thyroid gland is 30 gm, effective half-life is 4 days, and the measured thyroid uptake is 45%.

Calculation:

$$\text{Mass of Thyroid M} = 30 \text{ gm}$$

Initial amount of activity localized in the thyroid, $A_0$

$$= 0.45 \times 5000\ \mu\text{Ci}$$

$$= 2250\ \mu\text{Ci}\ (83{\cdot}25\ \text{MBq})$$

$$T_{\frac{1}{2}}(\text{eff}) = 4 \text{ days} = 4 \times 24 \text{ hr} = 96 \text{ hr}$$

Parameters such as $n_i$, $E_i$ and $\Delta_i$ for $^{131}I$, $\phi_i$ for thyroid for $^{131}I$ radiations and $\Sigma \Delta_i \phi_i$ are given in Table 7–6. Substituting these values in equation 6,

$$\text{D (thyroid)} = \frac{2250}{30} \times 1.44 \times 96 \times 0.433 \text{ rad}$$

$$= 4489 \text{ rad } (44{\cdot}89 \text{ Gy})$$

The important thing to note here is that if we had used only the particulate radiations of $^{131}I$ in our calculations, the radiation dose would have been

4043 rad, *i.e.*, 90% of the radiation dose to the thyroid by $^{131}I$ is delivered by particulate (mainly $\beta^-$) radiations.

## RADIATION DOSES IN ROUTINE IMAGING PROCEDURES

Table 7–7 lists the average radiation doses delivered to an adult patient from a number of radiopharmaceuticals commonly used in imaging various organs. Even though an internally administered radiopharmaceutical delivers a radiation dose to practically each and every organ in the body, only the total body and critical organ (the one which receives the highest radiation dose) radiation doses are used to illustrate the extent of the radiation doses routinely encountered in nuclear medicine. Their significance in terms of the risk involved in a patient is discussed in Chapter 15.

**Table 7–7. Radiation Doses in Common Imaging Procedures**

| | *Radioactivity Administered* | | *Radiation Dose (Total Body)* | | *Radiation Dose (Critical Organ)* | | |
|---|---|---|---|---|---|---|---|
| *Radiopharmaceutical* | *mCi* | *MBq* | *rad* | *mGy* | *rad* | *mGy* | *Critical Organ* |
| $^{99m}Tc$-labeled | | | | | | | |
| Pertechnetate | 10 | 370 | 0.15 | 1.5 | 2.5 | 25 | Stomach |
| Glucoheptonate | 20 | 740 | 0.15 | 1.5 | 4.5 | 45 | Bladder |
| Phosphate, etc. | 10 | 370 | 0.1 | 1 | 1.5 | 15 | Bladder |
| Sulphur colloid | 3 | 111 | 0.05 | 0.5 | 0.9 | 9 | Liver |
| MA Albumin | 3 | 111 | 0.05 | 0.5 | 1 | 10 | Lungs |
| DMSA | 6 | 222 | 0.1 | 1 | 4.0 | 40 | Kidneys |
| DTPA | 20 | 740 | 0.1 | 1.2 | 3.5 | 35 | Bladder |
| Red cells | 20 | 740 | 0.4 | 4 | 0.4 | 4 | Total body |
| Iron ascorbate | 2 | 74 | 0.008 | 0.08 | 1 | 10 | Kidneys |
| HIDA, etc. | 5 | 185 | 0.05 | 0.5 | 1.6 | 16 | Small Intestine |
| Mertiatide (Mag3) | 10 | 370 | 0.07 | 0.7 | 4.8 | 48 | Bladder |
| Sestamibi (Cardiolite) | 30 | 1110 | 0.5 | 5 | 5.4 | 54 | Upper large intestine |
| Teboroxime (Cardiotech) | 20 | 740 | 0.3 | 3 | 2.5 | 25 | Upper large intestine |
| exametazime (Ceretec) | 20 | 740 | 0.3 | 3 | 5.2 | 52 | lacrimal glands |
| $^{67}Ga$ Citrate | 5 | 185 | 1.3 | 13 | 4.5 | 45 | Lower large intestine |
| $^{111}In$DTPA (Cisternography) | 0.5 | 18.5 | 0.04 | 0.4 | 2.0 | 20 | Spinal chord |
| $^{131}I$-Iodide | 0.05 | 1.85 | 0.2 | 2 | 75 | 750 | Thyroid |
| $^{123}I$-Iodide | 0.1 | 3.7 | 0.004 | 0.04 | 2.2 | 22 | Thyroid |
| $^{123}I$-Iofetamine (SPECTamine) | 6 | 222 | 0.5 | 5 | 4.7 | 47 | Retina |
| $^{131}I$-Iodohippurate | 0.2 | 7.4 | 0.006 | 0.06 | 0.2 | 2 | Kidneys |
| $^{133}Xe$ | 10 | 370 | 0.001 | 0.01 | 0.3 | 3 | Lungs |
| $^{201}Tl$-Thallous Chloride | 3 | 111 | 0.7 | 7 | 4.5 | 45 | Kidneys |

## Radiation Dose to a Fetus

Since a fetus is more sensitive to radiation than an adult, it is not generally advisable to administer radiopharmaceuticals to a pregnant patient. However, there are occasions when, either because of strong medical reasons or inadvertently, a pregnant patient may receive radiopharmaceuticals. In these cases, it is essential to estimate the radiation dose to the fetus. The radiation dose to the fetus in these circumstances is derived from two sources: the radionuclide distributed in the mother's body and the radionuclide distributed in the fetal body due to the placental crossover. For most of the $^{99m}$Tc-labeled radiopharmaceuticals (except pertechnetate) there is little placental crossover and therefore most of the radiation dose is derived from distribution in the mother's body. In these cases, the estimated radiation dose to the fetus is between 30 to 40 mrad (300–400 mGy) from 10 mCi (370 MBq) of the administered radioactivity.

Radionuclides such as $^{131}$I, $^{67}$Ga and $^{201}$Tl, cross the placenta and therefore deliver a relatively large radiation dose.

## PROBLEMS

1. List the major factors on which the radiation dose from a radiopharmaceutical depends.
2. A $\beta$ emitter is distributed with uniform concentration in (a) a sphere, (b) a cylinder, and (c) a cube of the same volume. Is the radiation dose the same or different in the three cases?
3. What if the radionuclide in problem 2 is a pure $\gamma$ emitter?
4. On what factors does the absorbed fraction, $\phi$, depend?
5. How does the 'S' factor simplify the radiation dose calculation? What are its limitations? Can this method be used to calculate a radiation dose for a given patient?
6. A $\gamma$ emitter is localized in a patient's liver. Can it deliver a radiation dose to his or her brain, thyroid, lungs, spleen, testes, or ovaries? If yes, which of these organs will receive the maximum and minimum dose?

# 8

# Detection of High-Energy Radiation

When high-energy radiation interacts with matter, it produces certain physical or chemical changes in matter. These changes, which can be either transitory or permanent, are the basis for detection of high-energy radiation (referred to as radiation, for conciseness, in this chapter). However, these changes generally are too minute to be directly detected by our senses. Therefore, highly complex methods have been developed to detect radiation. Before describing the various detection methods, however, let us consider two questions.

## WHAT DO WE WANT TO KNOW ABOUT RADIATION?

Basically, we are interested in knowing one or more of the following:

***Simple Detection:*** Is radiation present? This question usually does not arise in nuclear medicine because one always administers a known radionuclide to the patient. In special circumstances, however, such as contamination of surroundings or personnel, one may be faced with this question.

***Quantity of Radiation:*** How much radiation is present? In any use of radiation in nuclear medicine, one must answer this basic question, if not in absolute terms then in relative terms (*i.e.*, with respect to a standard). In dynamic studies, this question is modified as: How much radiation is present at a given time? This requires measurement of radiation flux or counting rate as a function of time.

***Energy of the Radiation:*** In nuclear medicine, unlike nuclear physics, one is not interested in knowing the energy of the radiation *per se*. From practical considerations, however, it is helpful in two ways: (1) in discriminating against undesired events which may be identified by energy analysis; and (2) in identifying two radionuclides which may be used simultaneously.

***Nature of Radiation:*** In general, in nuclear medicine one knows the kind of radiation with which one is involved. However, again in cases of contamination, one may be faced with the task of identifying the nature of the radiation.

## WHAT MAKES ONE RADIATION DETECTOR BETTER THAN ANOTHER?

The answer to this complex question primarily depends on the use one wishes to make of the radiation detector. However, the following properties of a detector are of significance in nuclear medicine.

***Intrinsic Efficiency or Sensitivity:*** Intrinsic efficiency ($E_i$) of a detector is the measure of its ability to detect radiation and is generally defined as the ratio of the number of rays of a given radiation ($\alpha$, $\beta$ or $\gamma$) detected to the number of rays incident on the sensitive volume of the detector:

$$E_i = \frac{\text{Number of rays detected by the detector}}{\text{Number of rays incident on the sensitive volume of the detector}}$$

A value of 0.5 (50%) for the intrinsic efficiency means that only one-half of the rays incident on the sensitive volume of the detector are being detected and that the other half simply did not interact within the sensitive volume. The higher the intrinsic efficiency of a detector, the better it is for use in nuclear medicine. Intrinsic efficiency of a detector primarily depends upon the linear attenuation coefficient ($\mu$ linear) and the thickness of the sensitive volume.

***Dead Time or Resolving Time ($\tau$):*** Dead time or resolving time (for our purpose we shall consider them interchangeable, even though the purists distinguish between the two) is a measure of the ability of a detector to function accurately at high count rates or radiation flux (intensity). For any detector, there is a small but finite interval $\tau$, between the time when a ray interacts with a detector and the time when the detector responds and the event is recorded. This interval $\tau$, is known as "dead time" or resolving time of the detector. What happens if a second ray arrives and starts interacting with the detector while it is still processing the first ray? The response of a detector under such circumstance is broadly described under two categories: paralyzable and non-paralyzable. Both of these mechanisms may be operative in the same detector. In the paralyzable situation when the second ray arrives within the dead time ($\tau$) of the detector, the detector becomes insensitive for another time interval equal to $\tau$ from the time of arrival of the second ray. For example, if the dead time of a detector is 100 $\mu$sec and the second ray arrives after 30 $\mu$sec of the arrival of the first, the detector becomes insensitive for a total of 30 + 100 = 130 $\mu$sec. Provided of course, no third ray arrives within this interval. Because then, this interval will be extended again depending upon the time of arrival of the third ray and so on. Thus, depending on the count rate which determines the average time interval between the arrival of two successive rays, in the paralyzable case a number of rays can be lost without registering in the detector. In the non-

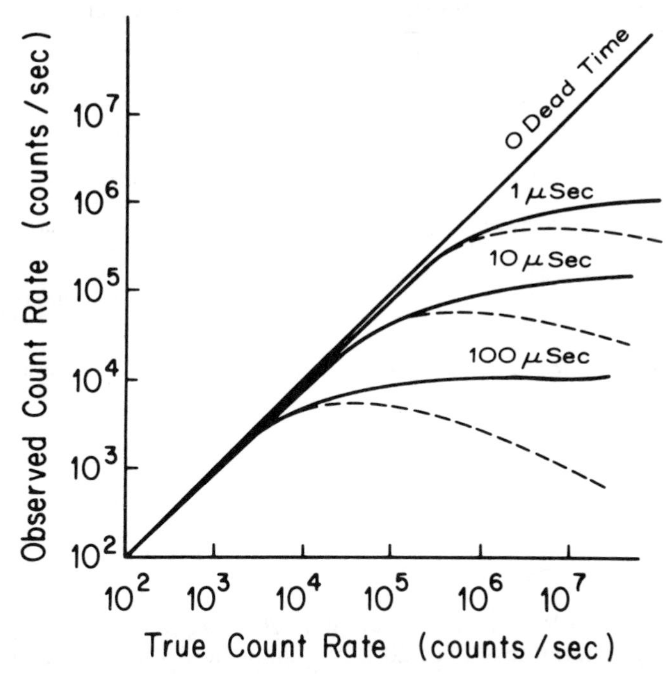


**Fig. 8–1.** Observed count rate vs true count rate for detectors with different dead times. Solid lines are for non-paralyzable detectors whereas broken lines are for paralyzable detectors. If both components, paralyzable and non-paralyzable, contribute to the dead time, the response will be in between the two extremes. Note that as the dead time of the detector increases, the usable count-rate range (linear portion) becomes smaller.

paralyzable case, the insensitive period ($\tau$), of the detector is not affected by the arrival of the second ray. The second ray is just lost. Figure 8–1 shows the effect of dead time on various count rates for a paralyzable and non-paralyzable detector. As can be seen, at high count rates, the detector response (observed counts) does not change with the increase in the true count rate for a non-paralyzable detector. For the paralyzable detector, the initial part of the response is the same but at high count rates, the response actually starts decreasing with the increase in the true count rate.

Ideally, the detector should have as short dead time as possible. For the count rates routinely encountered in nuclear medicine imaging, a system dead time of about 10 $\mu$sec is acceptable. However, for fast dynamic imaging of the heart where higher count rates are desirable, system dead time shorter than 10 $\mu$sec (2 to 3 $\mu$sec) are essential.

***Energy Discrimination Capability or Energy Resolution:*** Ability of a detector to distinguish between two radiations of different energies (*e.g.*, two $\gamma$-rays of different energies) is known as its energy discrimination capability. Full Width at Half Maximum (FWHM) which is commonly used as a measure of energy resolution represents the minimum difference necessary between the energies of two $\gamma$-rays, if they are to be identified as having different

**Table 8–1. Some Characteristics of Common $\gamma$-Ray Detectors**

| Detector | Intrinsic Efficiency | Dead Time ($\tau$) | Energy Discrimination | Uses in Nuclear Medicine |
|---|---|---|---|---|
| Ionization Chambers | Very Low | —* | None | Dose calibrators |
| Proportional Counters | Very Low | ~msec | Moderate | Rarely used |
| Geiger-Mueller Counters | Moderate | ~msec | None | Radiation survey work |
| NaI(Tl) Scintillation Counters | High | ~$\mu$sec | Moderate | Most widely used detector, well counters, rectilinear scanner, scintillation camera |
| Solid-State Ge(Li) Counters | Moderate | <1 $\mu$sec | Very Good | Neutron activation analyses |

* Cannot be used as a counter.

energies. If the FWHM of a detector is 20 keV, then two $\gamma$-rays with a difference in energy less than 20 keV cannot be distinguished by this detector. (See p. 116 for an explanation of the full meaning and scope of FWHM.) A lower value of FWHM indicates a better energy discrimination capability. Another way to look at FWHM is to consider it as an index of error in the measurement of energy of x or $\gamma$ ray by a detector. The lower this error is the more sensitive the detector will be in detecting small changes in the energy of x or $\gamma$ rays.

***Other Considerations:*** Since most detectors employ electronic components whose behavior quite often is susceptible to change with fluctuations in line voltage and environment temperature, detectors whose response is not appreciably affected by these fluctuations are preferred. Detectors should also be portable (depending on their desired use), simple to operate and, of course, inexpensive.

With this background in mind, I shall now discuss the various types of radiation detectors. These have been divided into three categories: gas-filled detectors, scintillation detectors, and solid-state detectors. Their main characteristics and uses are summarized in Table 8–1.

## GAS-FILLED DETECTORS

The first action of high-energy radiation on matter is the production of ionization. The measurement of the amount of ionization produced is the basis of gas-filled detectors. In general, it is not possible to measure the amount of ionization produced in matter except in gases and some solids known as semiconductors.

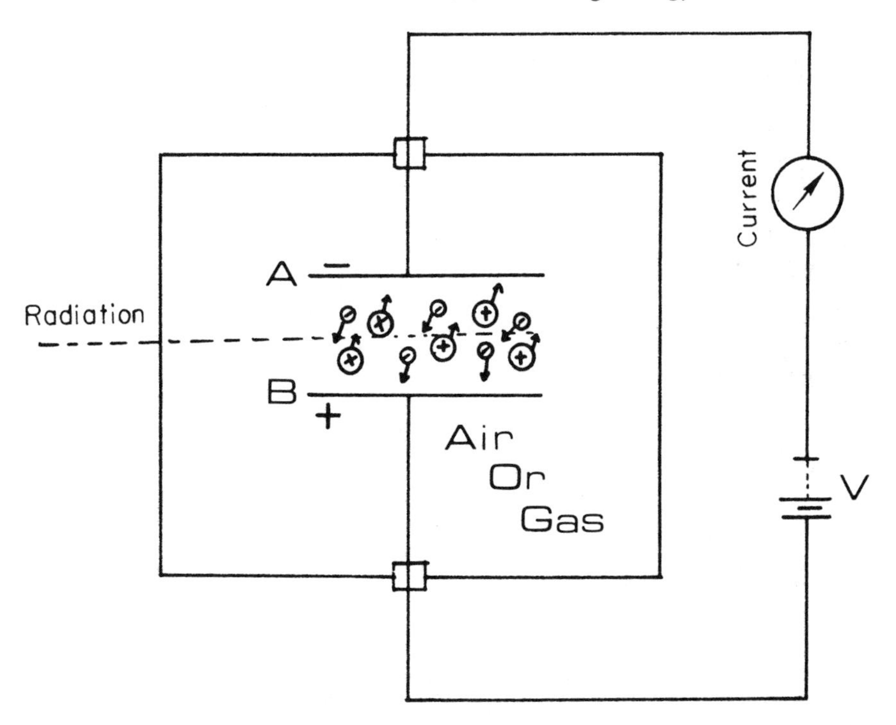


**Fig. 8–2.** Schematic of gas-filled radiation detector. A and B are two electrodes which collect the ion pairs produced in the gas by the passage of a radiation.

## Mechanism

To explain the mechanism of detection of gas-filled detectors let us consider what happens when ionization is produced in a gas which is enclosed between two electrodes with a voltage difference V, as shown in Fig. 8–2. When there is no voltage difference (V = 0) between the two electrodes, the ion pairs produced in the gas recombine to form neutral atoms or molecules, and no current flows in the circuit. However, under the influence of the electric field which exists between the two electrodes when V is greater than zero, some of the ion pairs reach the electrodes and a transient current is produced in the circuit. The amount of current produced depends on several factors, such as voltage difference V, separation distance of the two electrodes, nature of the gas, pressure and temperature of the gas, and geometry and shape of the electrodes. The most important parameter, however, is the voltage difference, V, between the two electrodes. The amount of the current produced in a typical gas-filled detector as a function of voltage V is shown in Fig. 8–3. As can be seen, the dependence of the current on voltage V is very complex. There are five distinct regions in this curve which need explanation.

In region I, the voltage V is low enough so that some of the ion pairs produced by radiation are still able to recombine and form neutral atoms or molecules. In other words, in this region there is an incomplete collection

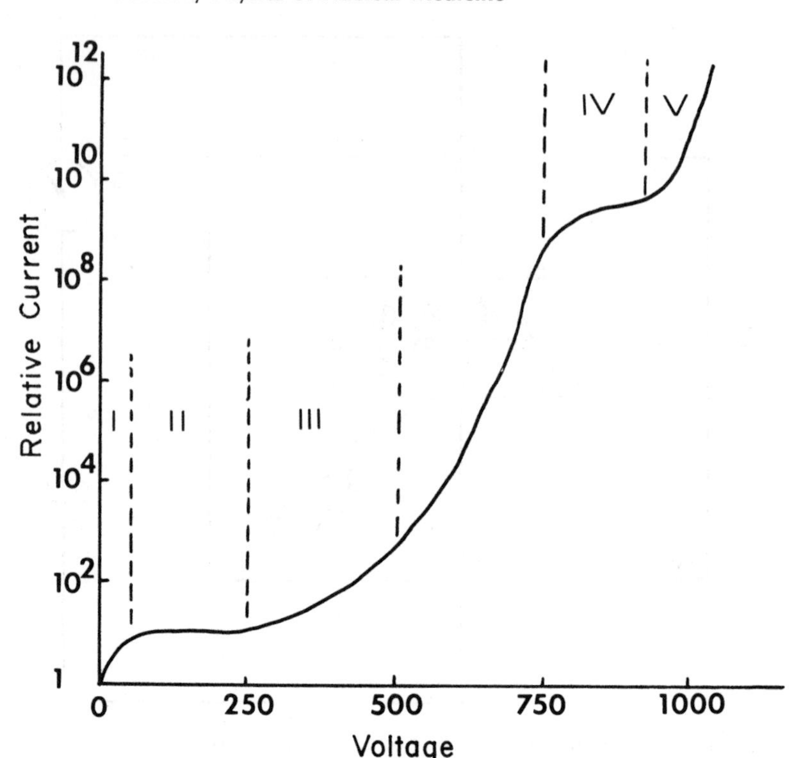


**Fig. 8–3.** Plot of current as a function of voltage (V) applied to the two electrodes A and B of Fig. 8–2. Origin of the five regions marked I, II, III, IV and V, is explained in the text. Ionization chambers operate in region II, proportional counters in region III and Geiger-Mueller counters in region IV. The unmarked region between regions III and IV is of no practical importance here.

of the ion pairs that are produced by the radiation. In region II, the voltage V is now sufficiently high to attract all the ion pairs produced by the radiation in the sensitive volume of the detector. In this region the amount of the current produced by an ionizing radiation does not change appreciably with change in voltage V. In region III, the voltage is high enough not only to attract all the ion pairs produced by the radiation but also to provide enough energy to some of the primary ion pairs to produce secondary ion pairs by collision with neutral atoms and molecules of the gas. In this region the amount of current varies significantly with change in voltage V.

In region IV, the voltage is so much higher that the primary ions produced by radiation acquire enough energy to produce a large number of secondary ion pairs and excitations, which then strike the metallic electrodes and neutral gas molecules to produce more ionization as well as excitation. The de-excitation of the molecules produces ultraviolet (UV) light, which then travels throughout the total gas volume producing more ionization. The end result of all this is a discharge (a turbulent state of ionization, excitations and UV light) in the total gas volume.

In this region, as in region II, the amount of current produced is more or less independent of the voltage V. In the last region, the voltage is so high that radiation is not necessary to produce discharge. Under the influence of such a high electric field, the electrons are pulled out from the atomic shells, the atoms and molecules become ionized, and a discharge may be established even without radiation.

### Ionization Chambers

Ionization chambers were one of the earliest types of gas-filled detectors used in radiation measurement. The operating voltage in these detectors lies in region II of Fig. 8–3, which is important because in this region small changes in voltage V do not change the current significantly. As a result, ionization chambers are highly stable and reliable. Their main disadvantages are their poor sensitivity (intrinsic efficiency) for x- or $\gamma$-rays and a lack of energy discrimination. Their principal use is in measurement of very high fluxes of radiation such as those encountered in diagnostic and therapeutic radiology. Recently, they have been used in nuclear medicine as "dose calibrators" to measure the amount of radioactivity in amounts of a few microcuries to curies.

***Dose Calibrator:*** Dose calibrators are, generally, cylindrically shaped ionization chambers, filled with argon at high pressure (~20 atmospheres). There is a small hole along the axis of the cylinder so that the radioactive source whose radioactivity is to be measured can be inserted close to the center of the ionization chamber (Fig. 8–4). This type of geometrical arrangement increases the overall sensitivity of a detector and is known as $4\pi$ arrangement (see also Chapter 9). The outside walls of the chamber are appropriately shielded to minimize the interference from radioactive sources outside the chamber.

The principle and operation of a dose calibrator are simple. The current produced in the ionization chamber by a given radioactive source in a given geometrical arrangement is directly proportional to the amount of the radioactivity of the source. However, different radionuclides with the same amount of radioactivity produce different amounts of current. Therefore, before an ionization chamber can be used as a dose calibrator, it has to be calibrated. The initial calibration is performed by measuring the amount of current produced for each millicurie of a given radionuclides. This process

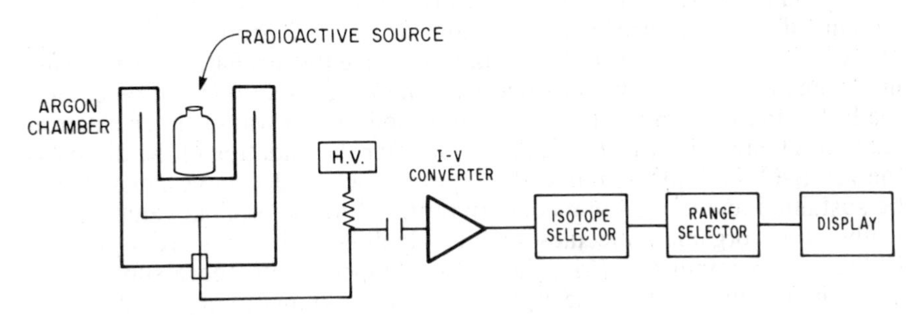


**Fig. 8–4.** A simplified schematic of a "dose calibrator."

is repeated for all the desired radionuclides. Sources of known radioactivity can be obtained from the National Bureau of Standards who determine the radioactivity of these sources accurately by other complex and time consuming methods. Once the calibration factors are known, then the unknown radioactivity of a given radionuclide is easily obtained by dividing the current produced by the unknown radioactivity with the calibration factor for that radionuclide. In commercial versions (Fig. 8–4) of dose calibrators, the ionization chamber is connected to a digital current meter and the calibration factors for a number of radionuclides are predetermined and set in such a way that the results of the unknown radioactivity are directly displayed in microcuries, millicuries or curies, etc. Remember though, when we press a button labeled $^{99m}$Tc all we are doing is calling the calibration factor of $^{99m}$Tc. We are not discriminating against other radionuclides. A $^{99m}$Tc source in a dose calibrator will display radioactivity at settings for other radionuclides as well, but these will not be correct readings of the $^{99m}$Tc radioactivity.

It must also be noted that the calibration factors are valid only for a given geometrical set up, source volume and source container. If the shape or type of the source container or the volume of the source are changed appreciably, the calibration factor will change and, therefore, should be measured again. To assure the proper operation of a dose calibrator, its linearity (*i.e.*, the current produced is proportional to the radioactivity) and stability with time should be periodically checked.

### Proportional Detectors (Counters)

These detectors operate in region III of Fig. 8–3, where there is a built-in amplification ($\sim 10^6$) of the primary ionization through the production of secondary ionization. As a result, a sufficient amount of current is produced by even a single ray; therefore, these detectors, unlike ionization chambers, can be used to count individual events. Proportional counters require a sufficient amount of expertise in their construction as well as their use. Their stability, in reference to time and voltage fluctuations, is not as good as that of ionization chambers. Proportional counters are rarely used in nuclear medicine.

### Geiger-Mueller (GM) Detectors (Counters)

The operating voltage of GM detectors is in the region IV of Fig. 8–3. The incoming particle in this case causes a discharge in the gas and the amount of the current produced is more or less independent of the energy of the radiation as well as of the voltage V. Once the discharge is established in the gas by a ray, how does one stop the process so that the detector is ready for the next ray? This is accomplished in two ways, either electronically or chemically. Electronically, soon after the initiation of the discharge the voltage V is suddenly reduced to a point at which discharge can no longer be sustained and is then immediately returned to the original value. Chemically, some organic molecules such as alcohols and ethers are introduced in the gas as impurities (*i.e.*, the gas is "doped" with these substances). The primary function of these organic impurities, known as chemical quenchers, is to absorb the UV light produced during the discharge, which then causes the discharge to stop. Absorption of energy by the quencher molecules leads to their dissociation. As a result and because there is only a small amount of quencher present, these types of GM tubes have a limited useful life. In modern versions of GM counters, quite often halogens or their organic compounds are used as chemical quenchers. These types of GM tubes have long lives because the halogens after dissociation by the absorption of kinetic energy or UV light can recombine to form the original molecule. Therefore, there is no depletion of halogens with usage, as happens with the organic type chemical quenchers. It takes approximately 50 to 200 $\mu$sec, either chemically or electronically, to quench the discharge. During this time the GM counter will not respond to another radiation; therefore, this is approximately the dead time of the detector. The maximum usable count rate for a typical GM counter is 50,000 counts/minute.

GM detectors are the most sensitive of the gas-filled detectors. They can be made in any shape and size, are easy to operate, and quite stable in reference to temperature and voltage fluctuations; however, they do not have any energy-discriminative capabilities. For the detection of $\beta$ rays, there is a small window made of thin metal either at the end or one side of the GM tube. For the detection of x- or $\gamma$-rays, this window is generally closed. Also, since the interaction of x- or $\gamma$-rays in the gas volume itself are minimal, the detection of these radiations is primarily through the photo- and compton electrons generated in the inner walls of the GM tube. The principal use of GM detectors in nuclear medicine is in radiation protection work for surveying the contamination of surroundings and personnel from x- and $\gamma$-rays emitting radionuclides.

## SCINTILLATION DETECTORS (COUNTERS)

A variety of substances, both organic and inorganic, scintillate (produce light) under the influence of high-energy radiation. This property is utilized for the detection of radiation by instruments known as scintillation detectors or counters. However, for the detection of the light generated in a material, it should be able to be transmitted out of the material itself. In liquids, it poses no serious problem. For solids, only single crystals or glasses can be used because in powders (microcrystals) light is absorbed and scattered at the boundaries of the microcrystals, thereby causing significant and variable loss of light before its detection.

A typical scintillation counter is shown schematically in Fig. 8–5. It consists of a scintillator, photomultiplier (PM) tube to detect the light, preamplifier, amplifier, and assorted electronics such as pulse-height selector (PHS) and a count-rate meter or scaler for automatic data collection and analysis. The following section outlines the mechanism of a typical scintillation counter used in the detection of $\gamma$-rays. Liquid scintillation counters used in the detection of high-energy charged particles (*i.e.*, $\alpha$ and $\beta$) are described in Chapter 9.

### Scintillator

The number of scintillating substances (scintillators) is large. Anthracene, naphthalene, various plastics, alkali halide crystals doped with small im-

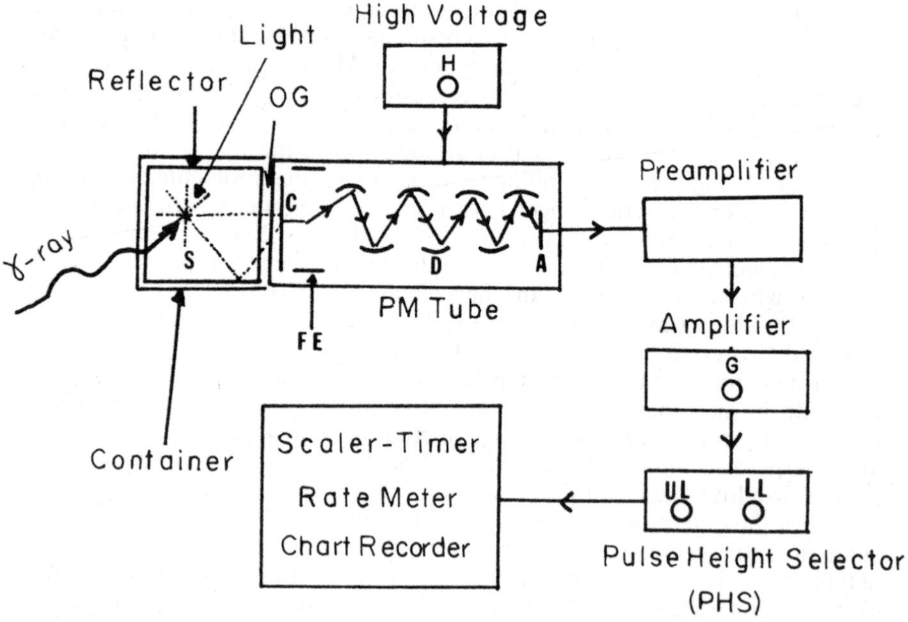


**Fig. 8–5.** Schematic of a scintillation counter. The four controls denoted by H (high voltage), G (gain of the amplifier), UL (upper level) and LL (lower level) are important in the proper operation of a scintillation counter. (In commercial detectors these may be denoted by various other names.) S stands for scintillator, OG for optical grease, C for photocathode, FE for focusing electrode, D for dynode and A for anode.

purities, lead tungstate, bismuth germanate (BGO), cesium fluoride (CsF), and aromatic compounds such as terphenyl, 2,5-diphenyloxazole (PPO) have all been used as scintillators.

Since the mechanism of light production in a substance under the influence of radiation is complex and not well understood, there are no theoretical laws to predict the behavior of a given substance in this regard. Briefly, a γ-ray loses its energy in a scintillator through the photoelectric, compton or pair-production mechanisms. The electrons thus generated then lose their energy within short distances through ionization and excitation of the scintillator molecules. The ion pairs thus produced then combine among themselves or with other atoms or molecules of the scintillator and produce certain excited states which, during their subsequent decay, emit light. The nature and quantity of these excited states determine the amount, color and phosphorescent decay time of the emitted light.

A scintillator is primarily characterized by (1) intrinsic efficiency, (2) amount of light produced per unit of absorbed energy (light conversion efficiency), and (3) the time in which emission of light takes place (phosphorescent decay time). The intrinsic efficiency of a scintillator for a γ-ray with a given energy depends on its linear attenuation coefficient, which in turn depends upon the atomic number and the density of the material (see Chapter 6). The amount of light produced per unit of absorbed energy which, in effect, determines the energy resolution of these detectors, and the phosphorescent decay time which is an important parameter for the dead time of these detectors, vary from substance to substance.

Of all the known scintillators, sodium iodide crystals doped with small amounts of thallium, NaI(Tl), are most widely used in nuclear medicine. Its moderate density (d = 3.67 gm/cm$^3$) and effective atomic number ($Z_{eff}$ = 450 make it very efficient for the detection of x- or $\gamma$-rays in the energy range of 30–500 keV. The amount of light produced per unit of absorbed energy in a NaI(Tl) crystal is also one of the highest, in spite of the fact that sodium iodide crystals without thallium doping do not produce much light. The presence of small amounts of thallium (1 part in $10^6$) enhances the light output by a factor of ten or more. The phosphorescent decay time, which eventually determines the dead time of a scintillation detector, is $\approx$0.25 $\mu$sec, and is adequate for most of the count rates presently encountered in nuclear medicine. In addition, the technology required to grow these crystals in large sizes and various shapes is well advanced, making their use economical.

Light in a scintillator is produced in a very small volume, mainly determined by the range of photoelectrons or compton recoil electrons produced in the scintillator. For energies of x- or $\gamma$-rays less than 1 MeV, this range does not exceed more than a millimeter in a NaI(Tl) crystal. From this small volume of production, light travels in all directions, as shown in Fig. 8–5. The direction of most of the light toward the PM tube is achieved by coating the outside surface of the scintillator, except the side facing the PM tube, with a light reflector such as magnesium oxide.

Sodium iodide crystals are hygroscopic and therefore have to be hermetically sealed, usually in thin-walled aluminum or steel containers. The coupling of the crystal assembly to the PM tube is also important because of possible light loss at the interface of the crystal and PM tube. This loss is generally minimized by the use of optical grease. Also, since the amount of light produced in these crystals is somewhat dependent upon ambient temperature, the operating temperature should not change appreciably.

These crystals should not be subjected to abrupt temperature changes (no more than 10°C/hour change); even when these are not in use (*e.g.*, during shipment or storage). Such changes can produce severe mechanical stresses in the crystal causing them to crack.

### Associated Electronics

***Photomultiplier (PM) Tube:*** The amount of light produced in NaI(Tl) crystals is quite small relative to that which the human eye can easily detect. Even if one could observe each scintillation occurring in the scintillator with the naked eye, it is not very practical to count light flashes in this way for long periods of time. A PM tube is a light-sensitive device which converts light into measurable electronic pulses. It consists of a photocathode facing the window through which light enters, a series of metallic electrodes known as dynodes arranged in a special geometric pattern, and an anode—all enclosed *in vacuo* in a glass tube. When the light photon hits the photocathode it produces an electron of low energy (0.1 to 1 eV) through photoelectric interaction. This photoelectron is then accelerated toward a dynode by the application of a voltage (between 50 and 100 volts) to that dynode. As a result of this acceleration, the electron acquires sufficient kinetic energy (50 to 100 eV) to produce a number of secondary electrons when it collides with the dynode. The number of secondary electrons produced varies between 1 and 10. These secondary electrons are then accelerated toward a second dynode where a similar multiplication in the number of electrons occurs. Eventually, at the last dynode (generally the tenth), there are between $10^5$ to $10^8$ electrons for each of the photoelectrons produced. These electrons generate a current pulse of a few microamperes in amplitude and about a microsecond in duration at the anode.

In a typical PM tube the voltage to the different dynodes is supplied from a single high voltage source (500 to 1500 volts) by employing a voltage divider. The gain of a PM tube (the multiplication of electrons) strongly depends on the voltage applied to a dynode and, therefore, on the high voltage itself.

***Preamplifier:*** The electric pulses arriving at the anode of a PM tube are very small (in millivolts and microamperes) and, therefore, require amplification to several volts before they can be further analyzed or processed. The output of a PM tube, however, cannot be directly fed into an amplifier because of the wide differences in the output impedance (an important parameter of electronic circuits) of a PM tube and the input impedance of an amplifier which results in signal distortion and attenuation. A preamplifier is a device which primarily solves the problem of such impedance mismatch. It is always located as close to the PM tube as possible because long connecting cables at this stage attenuate the signal significantly. The output of the preamplifier, however, can be easily transmitted through cables over long distance.

***Linear Amplifier:*** A linear amplifier amplifies the electric pulses arriving from a preamplifier and processes them in such a way that they are suitable for the subsequent data analyzing equipment. The ratio between the amplitudes of the outgoing and incoming pulse is known as amplifier gain. If desired, the amplification can be changed by suitably adjusting the gain control, specifically provided for this purpose on an amplifier.

***Pulse-Height Selector (PHS):*** A pulse-height selector is an electronic device that selects only those pulses whose voltage lies in a preselected range and rejects those pulses whose voltage lies outside this range. A PHS is provided with two controls known as lower level and upper level or, sometimes, lower level and window. These two controls determine the desired range of the pulses to be selected. For example, if pulses with a voltage varying from 1 to 10 volts are being produced by the linear amplifier in response to a given x- or $\gamma$-ray, in order to select only those pulses whose voltage lies between 5 and 6 volts it is necessary to set the lower level of PHS at 5 volts and the upper level at 6 volts. The PHS will then produce a pulse only when the incoming pulse lies in this range. In a PHS with lower-level and window controls, these would be set at 5 volts and 1 volt, respectively.

In some nuclear medicine instrumentation, there is another control known as integral counting. On this setting, only the lower level control is operative and all pulses higher than the voltage indicated by the lower level setting will be counted. Also, in many instruments, the two controls, lower level and upper level (or window) have been replaced by center line or peak voltage and % window. In this terminology, if we wish to select pulses that fall between 5 and 6 volts, then centerline or peak voltage will be set at 5.5 volt (middle of the range) and % window $= \frac{\text{desired window width}}{\text{peak voltage}} \times 100$ which in this case will be $\frac{1 \text{ volt}}{5.5 \text{ volt}} \times 100 \simeq 18\%$. Alternatively, a 20% window at 1.40 peak voltage will be $0.2 \times 1.40 = 0.28$ volt wide. In this case, the window is set symmetrically on each side of the peak voltage. Thus, the pulses selected will have their voltages from $1.40 - \frac{0.28}{2}$ to $1.40 + \frac{0.28}{2}$ or 1.26 volt to 1.54 volt.

***Multichannel Analyzer:*** A PHS is a single-channel analyzer because one can select only one range. In a multichannel analyzer, there are many PHS's which can simultaneously separate the pulses into a multiple number of preselected voltage ranges.

***Scaler and Timer:*** These devices comprise an electronic counter used for counting the pulses coming from an amplifier or PHS either for a specific time period or until a predetermined number of counts have been collected. In either case, both the number of pulses collected and the time during which they are collected are obtained.

***Ratemeter:*** This is a device which, instead of giving the number of counts and time period in which those counts were collected separately, directly yields the count rate (counts/min). A ratemeter always has associated with it a parameter known as the time constant. The time constant of a ratemeter determines the statistical accuracy of the count rate as well as its response or reaction time to the changes in the count rate (Fig. 8–6). The time constant of a ratemeter plays an important role in recording the dynamic studies. Care should be exercised in choosing an optimum value of the time constant which will provide, on one hand, a good statistical accuracy and, on the other, a faithful reproduction of the dynamic phenomena.

### Response to Monochromatic (Single-Energy) $\gamma$-Rays

Let us now review the steps leading to the production of the voltage pulse in a scintillation detector. A $\gamma$-ray interacts within the crystal volume through the photoelectric, compton or pair-production mechanism. In each case, either a high-energy electron or electron-positron pair is produced which deposits all its energy within a small volume of the site of production. Some of the energy deposited by the electron or electron-positron pair is converted into light. Light triggers an electric pulse in the PM tube which is subsequently amplified by the linear amplifier.

The amplitude (height or voltage, V) of the pulse coming out of the amplifier is directly proportional to the amount of energy deposited by a $\gamma$-ray

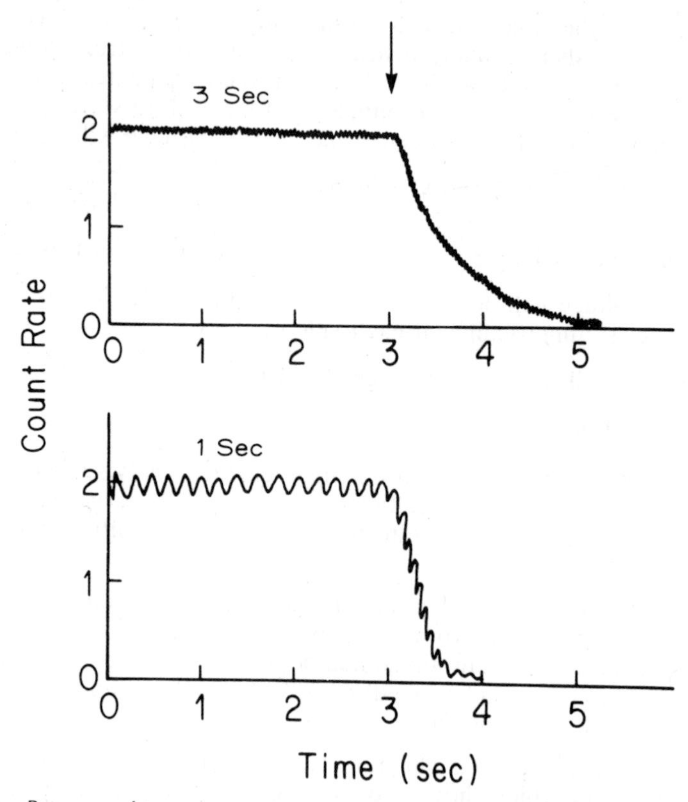


**Fig. 8–6.** Response of two ratemeters with 1- and 3-sec time constants, respectively, as traced on a chart recorder when a radionuclide source is suddenly moved a long distance from the detector. The arrow (top) represents the time at which the radioactive source was removed. The statistical fluctuations for the 1-sec time constant (measured by the magnitude of the waves) are larger than those for the 3-sec time constant; the response time (measured by how fast the count rate drops to zero once the source has been removed) for the 1-sec time constant is much shorter than for the 3-sec time constant.

($E_d$), the light conversion efficiency of the scintillator (N), the efficiency of photocathode (E), the PM-tube gain ($G_{pm}$) and the amplifier gain ($G_{amp}$). In other words

$$V = \text{constant} \cdot E_d \cdot N \cdot E \cdot G_{pm} \cdot G_{amp},$$

where N, E, $G_{pm}$ and $G_{amp}$ are characteristic of a given scintillation counter.

In the above equation, if one keeps the PM-tube gain ($G_{pm}$) and amplifier gain ($G_{amp}$) constant, then the pulse height V produced in a given scintillator detector depends linearly on the energy deposited ($E_d$) by the $\gamma$-ray. As $E_d$ increases so does the pulse amplitude V. This, in essence, is the source of the energy-measuring capability of a scintillation detector.

A $\gamma$-ray of energy, $E_\gamma$, however, does not deposit the same amount of energy ($E_d$) each time it interacts with a crystal. The $E_d$ depends on the type of interaction (*e.g.*, photoelectric, compton or pair production). In photoelectric events all the $\gamma$-ray energy is deposited in the crystal, *i.e.*, $E_d = E_\gamma$. In compton and pair-production events, $E_d$ is always less than $E_\gamma$ and there is no simple relationship which relates $E_\gamma$ to the pulse amplitude V. Therefore, in order to perform energy analysis with a scintillation detector, one has to select only the photoelectric events. This is done with the help of a PHS.

In Fig. 8–7 we have shown the pulse amplitude distribution produced by a NaI(Tl) detector for a monochromatic $\gamma$-ray whose energy is below 1 MeV. In this case only photoelectric and compton interactions are important. The pulse amplitude distribution (also known as spectrum) is divided into two regions, *a* and *b*.

Region *a* is known as the compton plateau, and all the pulses in this region

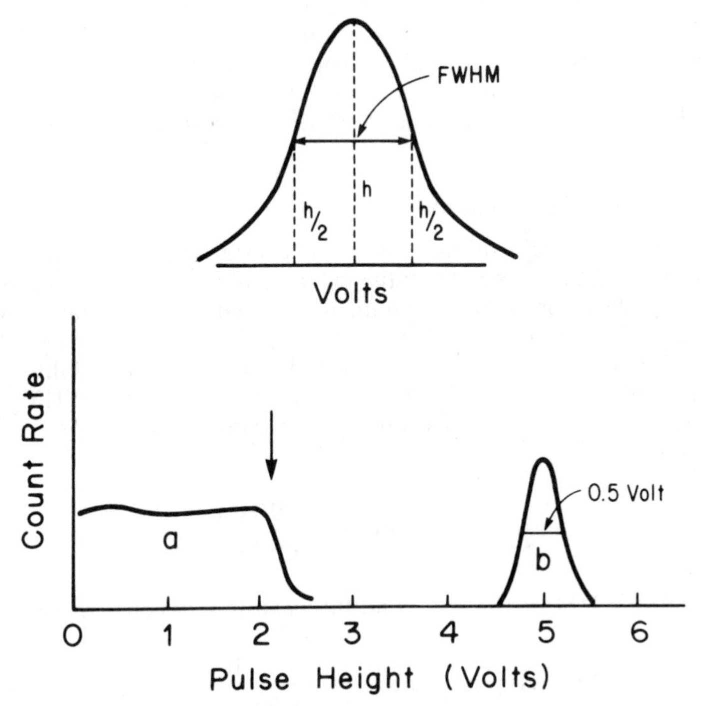


**Fig. 8–7.** Pulse-height distribution produced by a NaI(Tl) scintillation detector in response to $\gamma$-rays of a single energy. The distribution is divided into two distinct areas, a and *b*. Area a (compton plateau) results primarily because of the compton interaction of $\gamma$-rays in the NaI(Tl) crystal. The edge of the compton plateau (indicated by arrow) results when a $\gamma$-ray transfers maximum energy possible in a compton interaction to an electron. Area *b* (photopeak) is produced when $\gamma$-rays interact in the crystal via photoelectric effect. The narrowness or width of the photopeak is indicative of the energy-resolving capability of a detector. Quantitatively, energy-resolving capability is measured by FWHM (Full Width at Half Maximum). FWHM is shown diagrammatically in the inset (top). Energy resolution is then measured by $\frac{\text{FWHM} \times 100}{\text{Peak Voltage}}$. The lower this value, the better the energy resolution.

originate as a result of compton interaction. Depending upon the amount of energy given to the compton recoil electron, pulses of different amplitudes are produced. Since, in compton interaction, a $\gamma$-ray does not transfer all its energy to an electron, the maximum transfer of energy to the electron takes place when the $\gamma$-ray is back-scattered (see Chapter 6, p. 82). When such an event occurs it corresponds to the compton edge indicated by the arrow in Fig. 8–7. Region *b*, a bell-shaped curve, is known as a photopeak. Almost all the pulses in this region are primarily produced as a result of photoelectric interaction.

Some pulses in this region may also arise as a result of multiple compton interactions or a compton interaction and subsequent absorption of the scattered gamma ray through photoelectric interaction. In either event, total energy of the gamma ray has to be deposited in the crystal. Therefore, the photopeak is, sometimes, called total absorption peak.

Since these two regions are generally separated by a valley, the photoelectric events can be easily selected by using a PHS. The lower and upper levels of a PHS are so set as to just encompass the photopeak. In Fig. 8–7 these may be set at 4.5 volts and 5.5 volts, respectively.

A good rule of thumb to select the window width for any gamma ray is for it to be about 2 times the energy resolution (discussed below) for that gamma ray peak.

### FWHM and Energy Resolution

The processes of light production in the crystal and the multiplication of electrons in a PM tube are both statistical in nature. As a result, the response of a NaI(Tl) detector is not constant (or a single fixed voltage) even for photoelectric events, but is a bell-shaped distribution (gaussian). The narrower this distribution, the better energy-discrimination capability of a detector. The full width of this distribution at half maximum (FWHM) is used as a measure of the energy-discriminating capabilities or energy resolution of a detector. Generally it is expressed as

$$\%\text{ energy resolution} = \frac{\text{FWHM} \times 100}{\text{PEAK voltage}}$$

In Fig. 8–6, FWHM = 0.5 volts and the peak is at 5 volts; therefore

$$\%\text{ energy resolution} = \frac{0.5 \times 100}{5.0} = 10\%$$

Since the amount of light produced in the crystal for the total absorption peak increases linearly with the gamma ray energy, the energy resolution is dependent on the energy of gamma ray. The higher the energy of the gamma ray, the better the energy resolution, though these are not linearly related. Another factor on which energy resolution depends to a slight extent is the size and shape of the sodium iodide crystal. For a well type crystal, energy resolution is poorer than a cylindrical type crystal.

In a good NaI(Tl) detector system, the best obtainable percent energy resolution for a 662 keV gamma ray of $^{137}$Cs is between 8 and 10%. The percent energy resolution for 140 keV gamma rays of $^{99m}$Tc ranges from 11 to 14%.

Periodic measurements of the energy resolution can be used to diagnose the slow deterioration of the crystal due to moisture leak in the crystal housing and/or the PM tube weakening. Energy resolution gets worse with time as a result of any of these problems.

### Variations Resulting from Change in Amplifier Gain or High Voltage

Small increases or decreases in PM tube high-voltage or amplifier gain proportionally increase or decrease the pulse amplitude (voltage), as shown in Fig. 8–8. If one has already set the lower and upper level of a PHS to encompass the photopeak, then, a change in amplifier gain or high voltage can move the photopeak outside the window or the selected range in a PHS. Therefore, it is important that the high voltage and the amplifier gain is stable with time, ambient temperature changes, and line-voltage fluctuations.

### Energy Calibration

Since by arbitrarily increasing the high voltage and/or amplifier gain one can produce a pulse of any desired amplitude for a fixed-energy $\gamma$-ray, how

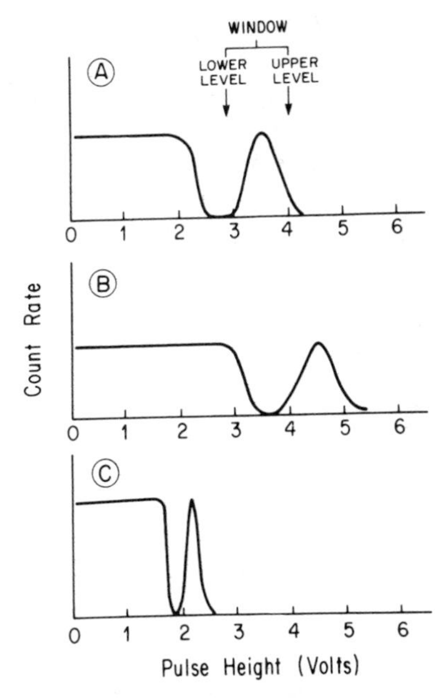


**Fig. 8–8.** Effect of the change in high voltage (or amplifier gain) on pulse-height distribution. An increase in high voltage from that of (A) produces a pulse-height distribution shown in (B), whereas a decrease in high voltage produces a pulse-height distribution shown in (C). In these extreme cases (B and C), a photopeak within the selected range of a PHS moves out of the selected range (window) completely, thereby reducing the count rate drastically. In practice, small variations in the high voltage or amplifier gain can result in appreciable changes in the count rate obtained with a preset window. Therefore, the high voltage and the gain of the amplifier should be stable.

does one calibrate the pulse amplitude to the γ-ray energy? This is done by using radionuclide sources which emit a single, known energy γ-ray and are long-lived, *e.g.*, $^{137}$Cs which emits a γ-ray having an energy of 662 keV. A calibration commonly employed in nuclear medicine is such that 100 keV energy deposited in the crystal produces a pulse of 1.0 volt amplitude. To achieve this calibration one adjusts the PM-tube high voltage in such a manner that the $^{137}$Cs photopeak is centered at 6.62 volts. Once this calibration has been performed, neither the amplifier gain nor the high voltage should be changed. Also, as most electronic instruments drift in gain, the energy calibration should be checked frequently.

### Response to γ-Rays of Two Energies

In this case, the pulse-height distribution is the sum of the individual pulse-height distribution, of two γ-rays (γ1) and (γ2), as shown in Fig. 8–9. The photopeak of the higher energy γ-ray (γ2) is still isolated, and the lower energy γ-ray (γ1) causes no interference in its detection. However, the photopeak of γ1 lies in the compton plateau region of γ2 and there is no way to avoid the compton pulses of γ2 from being counted in the PHS window

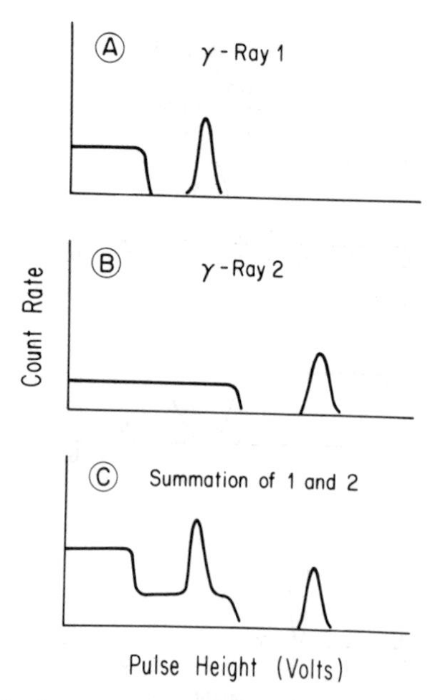


**Fig. 8–9.** Pulse-height distribution produced in response to γ-rays consisting of two distinct energies, (1) and (2). As shown in (C), this is an exact sum of the pulse-height distributions produced by γ-rays of energy (1) and energy (2) independently, as shown in (A) and (B), respectively. Low-energy γ-rays do not produce pulses with enough height to cause any interference in the photopeak of the higher energy γ-rays. However, high-energy γ-rays produce pulses with a height corresponding to the photopeak pulses of lower energy γ-rays. Therefore, a correction has to be applied whenever a low-energy γ-ray radionuclide is being counted in the presence of a higher energy γ-ray radionuclide. This is not so in the opposite case.

for the photopeak of $\gamma 1$. The compton contribution of these pulses to $\gamma 1$, however, can be determined by using a source which emits a single $\gamma$-ray with energy very close to that of $\gamma 2$. Using this source, the contribution of $\gamma 2$ in the photopeak window of $\gamma 1$ is determined as a percentage of the photopeak of $\gamma 2$. Then, knowing the counts in the photopeak of $\gamma$ in an unknown sample, one can determine the number of compton counts which will be produced in the photopeak of $\gamma 1$.

### Secondary Peaks

Quite often, besides the photopeak and compton plateau, additional but much smaller peaks can be recognized in a gamma-ray spectrum. These should not be confused with the presence of some weaker gamma rays in the radionuclide emissions. Rather, these have well defined characteristics and causes as described below.

***K-Escape Peak:*** This peak, mostly observed when the energy of an interacting gamma ray lies between 50 and 150 keV, always lies about 28 keV below the photopeak of the gamma ray. In this energy range (50–150 keV), the majority of the x or $\gamma$ interactions are through the photoelectric effect in the K shell of iodine atoms and most of these interactions are close to the crystal surface (within a few millimeters). Therefore, the K x-ray produced when the vacancy in the K shell of the iodine atoms is filled, has a finite chance of escaping from the crystal without depositing its energy. Thus, in these cases, the total energy deposited by the gamma ray even when the interaction was through photoelectric effect is not the total energy of the gamma ray, $E\gamma$, but equals $E\gamma - E_{k\ x\text{-ray}}$. The energy of K x-ray for iodine atom is about 28 keV, hence the escape peak is 28 keV below the photopeak of the gamma ray.

***Summation Peak:*** With radionuclides emitting more than one-energy gamma ray, sometimes additional peaks corresponding to the sum of the individual gamma ray energies are encountered even though no gamma ray with such energy is emitted by the radionuclide. As the name implies, these peaks originate because of the summation of the energies of two different gamma rays in the crystal. Summation of energies occurs only when two gamma rays simultaneously or almost simultaneously (within 0.25 $\mu$sec) interact with the crystal. Thus, summation peaks are more pronounced for radionuclides that emit several gamma rays within one $\mu$sec or less time of each other and for well-type scintillation detectors (Chapter 9). Some examples of this type of radionuclide are $^{125}I$ (28 K x-ray followed by another 28 K x-ray) and $^{111}In$ (173 and 247 keV gamma rays). If a spectrum is plotted for these radionuclides employing a well-type scintillation detector besides the 28 keV x-ray of $^{125}I$, another peak at $28 + 28 = 56$ keV is also observed, even though $^{125}I$ does not emit any x or $\gamma$ radiation with 56 keV energy. Similarly in the case of $^{111}In$, a summation peak at $173 + 247 = 420$ keV is observed, even though $^{111}In$ does not emit a gamma ray with this energy. For this type of summation, the two gamma rays do not have to interact through the photoelectric effect only. The interaction may occur through compton interaction as well. Then the summated pulse which is also called "pile up" pulse will be equal to the sum of the energy deposited by each gamma ray. At high count rates "pile up" pulses result accidentally, even when the radionuclide emits one-energy gamma ray only.

***Back Scattered Peak and Lead X-Ray Peak:*** These peaks originate from the environment or surroundings of the NaI(Tl) detectors. Since to reduce the background radiation, the detectors are generally shielded by lead from all sides except one, the gamma rays interacting with the lead shield produce scattered gamma rays as well as lead K x-rays of about 80 keV. Some of these secondary radiations can be detected by the crystal. Of the scattered gamma rays from the shield, only those which are scattered more or less in the backward direction have a chance to interact with the crystal. The energy of these back scattered gamma rays is determined by the expression given on p. 82. The back scattered gamma rays are not of a single energy but are distributed in a broad range. As a rule, the sum of the back-scattered peak

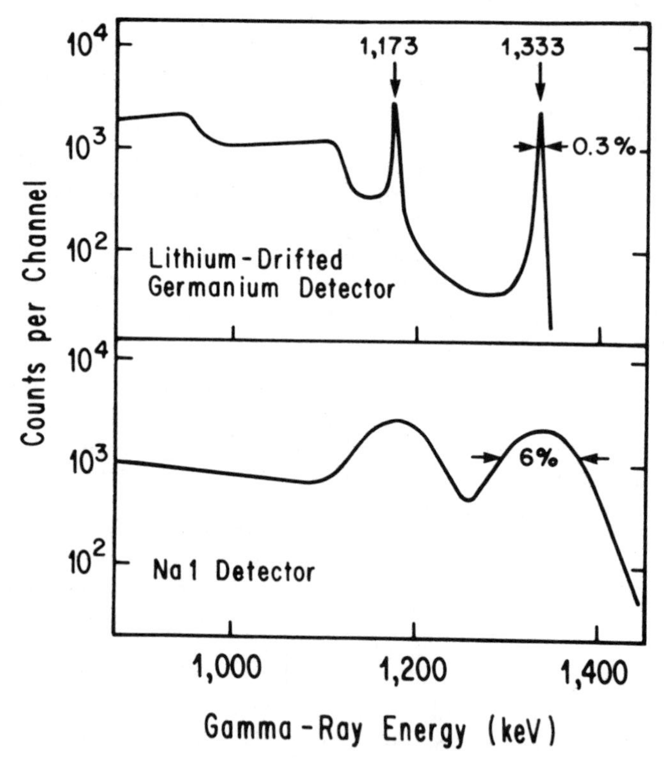


**Fig. 8–10.** Gamma ray spectrum of a $^{60}$Co source (two gamma rays with 1173 and 1333 keV energy each) taken with a Ge(Li) semiconductor detector and a NaI(Tl) detector respectively. With the Ge(Li) detector, the two photopeaks are narrow and clearly resolved. With the NaI(Tl) detector, the two photopeaks are broad and barely resolvable. The energy resolution for 1333 keV gamma ray is 0.3% for the Ge(Li) detector and 6% for the NaI(Tl) detector. If a third gamma ray with an intermediate energy of about 1250 keV was also present in the $^{60}$Co emissions, a NaI(Tl) detector would have been unable to resolve the three gamma rays, whereas a Ge(Li) detector would have identified them clearly.

energy and the compton-edge energy in the gamma ray spectrum of a monochromatic gamma ray equals to the photopeak energy.

## SEMICONDUCTOR DETECTORS

These newly developed detectors have revolutionized nuclear physics. It is not possible, generally, to measure the amount of ion pairs produced by ionizing radiation in a solid, as in a gas. However, under certain conditions, for a class of substances known as semiconductors, it is possible to measure the amount of such ion pairs produced. The two most common semiconductor detectors use germanium and silicon crystals doped with a small amount of lithium, [Ge(Li) and Si(Li)]. Ge(Li) detectors are used for the detection of x- or $\gamma$-rays, whereas Si(Li) detectors are primarily used for the detection of corpuscular radiation.

The principal advantage of a Ge(Li) detector over a scintillation detector lies in its excellent energy resolution. A typical Ge(Li) detector can easily yield a 1% energy resolution for $^{137}Cs$ $\gamma$-rays compared to 10% using a NaI(Tl) detector. The main drawbacks of Ge(Li) detectors, severely limiting their application in nuclear medicine, are their low sensitivity, the necessity of maintaining them at a low temperature of 77°K (room temperature is about 300°K), and their unavailability in larger sizes.

A spectrum produced by a $^{60}Co$ source and obtained with a Ge(Li) detector is shown in Fig. 8–10, where it is also compared with a NaI(Tl) detector.

## PROBLEMS

1. List the factors on which the intrinsic efficiency of a detector depends. Why is this important in nuclear medicine?
2. Is the dead time of a detector related to the maximum detectable count rate directly, inversely, or with some other power?
3. Which of the detectors discussed here is appropriate for the following applications? 1. For surveying the extent of contamination when a technologist, during an injection of a radiopharmaceutical, has spilled some radioactivity on the patient's bed as well as his own clothing. 2. To determine a small amount (0.01 microcurie) of radioactivity in a sample. 3. To determine large amounts of radioactivities such as are to be given to patients. 4. To characterize different radionuclides in a mixture of radionuclides.
4. List the properties of the NaI(Tl) detector that make it so attractive for use in nuclear medicine.
5. What is the function of a pulse height selector? What effect will opening the window wider have on the observed count rate?
6. Three NaI(Tl) detectors have FWHM of 18, 20, and 22 keV for 140 keV $\gamma$ ray. What is the % energy resolution of these detectors and which one should you select?

# 9

# In-Vitro Radiation Detection

In many nuclear medicine procedures, such as the Schilling test, blood volume determinations, protein and fat absorption studies, ferrokinetics and various radioimmunoassays, it is necessary to determine the amount of radioactivity in a given sample as compared to a standard. In some of these tests (*e.g.*, Schilling test), a radioactive substance is administered to a patient; in others, such as various radioimmunoassays, no radioactive compound is given to the patient. In both cases, however, the choice of the radiation detector and the geometric set-up for the measurement of a given radioactive sample is of great importance and is primarily dictated by the overall efficiency (sensitivity) of a particular set-up. An increased overall efficiency in these studies allows one either to reduce the radiation dose to a patient or reduce the count time for a given standard deviation error.

This chapter describes the various techniques employed for the in-vitro detection of the radioactivity of both $\beta$-ray and $\gamma$-ray emitting radionuclides.

## OVERALL EFFICIENCY (E)

The overall efficiency (E) of a given device for the measurement of radioactivity depends on the geometric efficiency of the physical set-up ($E_g$) and the intrinsic efficiency ($E_i$) of the detector. Thus, the overall efficiency can be expressed as:

$$E = E_g \times E_i$$

***Intrinsic Efficiency:*** The intrinsic efficiency ($E_i$) (Chapter 8) is defined as:

$$E_i = \frac{\text{Number of radiations detected by the detector}}{\text{Number of radiations incident on the sensitive volume of the detector.}}$$

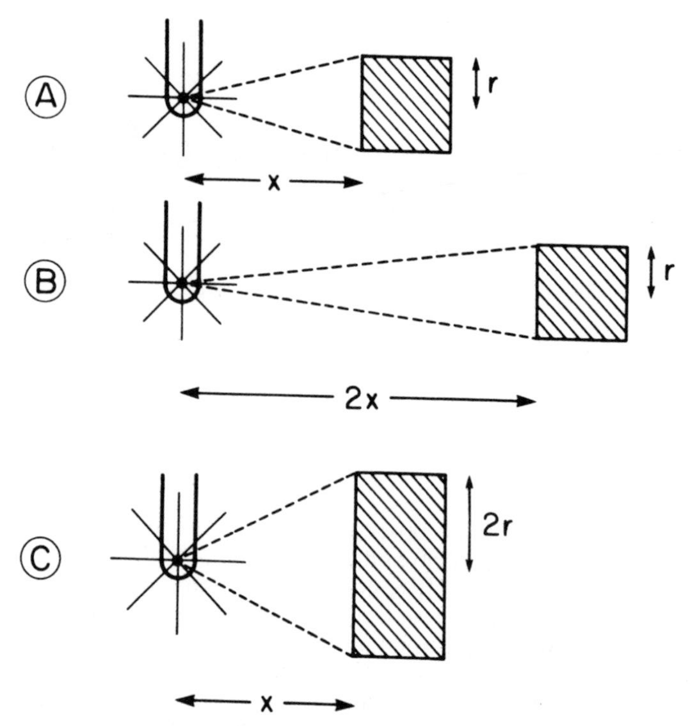


**Fig. 9–1.** Variations in geometric efficiency of a single radiation detector. A radioactive source emits radiations in all directions. Only those radiations which travel in the direction enclosed by the dotted lines reach the detector. The geometric efficiency, which is a measure of the number of $\gamma$-rays that reach the detector, depends on the distance, x, between the source and the detector, and the radius of the detector as $\frac{1}{x^2}$ and $r^2$. Thus, the geometric efficiency of arrangement (B) is one fourth of arrangement (A); the geometric efficiency of arrangement (C) is four times that of arrangement (A). In such an arrangement, however, geometric efficiency can never be higher than 50%.

It depends primarily on the linear attenuation coefficient, $\mu$(linear), of the material of which a detector is made.

In a scintillation counter, when one restricts oneself to the photopeak counts only, the resulting intrinsic efficiency is known as the photopeak efficiency.

***Geometric Efficiency:*** Geometric efficiency ($E_g$) can be understood as follows: Let a small sample of radioactive compound be located at a distance from a detector with a cross-sectional radius, r, as shown in Fig. 9–1A. Since there is no preferred direction for the emission of radiation from a radionuclide, the radiations ($\alpha$, $\beta$ or $\gamma$) from a radioactive sample will be emitted with equal probability in all directions. Therefore, out of the total number of nuclear radiations emitted in all directions only a fraction of these will be incident on the sensitive volume of the detector (Fig. 9–1A). Geometric efficiency, $E_g$, then, is defined as the ratio of the number of radiations incident on the detector from a radioactive sample to the total number of radiations emitted by the sample, or

$$E_g = \frac{\text{Number of radiations incident on detector}}{\text{Number of the total radiations emitted by source}}$$

The geometric efficiency, $E_g$, for the device shown in Figure 9–1A, depends on two factors: (1) the distance, x, between the source and the detector; and (2) the radius, r, of the cross-sectional area of the detector. For distances which are sufficiently larger than the radius of the detector (*i.e.*, $x \gg r$) the geometric efficiency varies as $\frac{1}{x^2}$ and $r^2$, respectively, with the distance x and the radius r. That is, if one increases the distance between the sample and the detector twofold, the geometric efficiency is reduced fourfold (Fig. 9–1B). If the radius of the detector is doubled, the geometric efficiency will be increased fourfold (Fig. 9–1C).

When the sample is very close to the detector (*i.e.*, $x \ll r$), the above relationship $\left(E_g \alpha \frac{1}{x^2}\right)$ does not hold true. The exact relationship between $E_g$ and the distance x, in this case, is of no practical concern to us here except to know that the maximum obtainable geometric efficiency is achieved when $x = 0$ (*i.e.*, the source is directly against the face of the detector). In this case, $E_g$ approaches 50%, because even in this situation one-half of the radiations are emitted away from the detector.

To increase $E_g$ to more than 50%, one either has to use more than one detector or somehow surround the sample from all sides by the detector. The second approach is used in well-type NaI(Tl) scintillation detectors (well counters) and liquid scintillation detectors described below.

## WELL-TYPE NaI(Tl) SCINTILLATION DETECTORS (WELL COUNTERS)

In terms of its operation and associated electronics, the well-type NaI(Tl) scintillation detector (well counter) is exactly the same as the NaI(Tl) scintillation detector described in Chapter 8, except that, in a well counter, a small cylindrical hole is constructed in the NaI(Tl) crystal in order to allow a radioactive sample to be positioned very close to the center of the crystal (Fig. 9–2). In this type of detector, only a small fraction (<5%) of the radiations emitted by the sample escapes from the sensitive volume of the crystal and, therefore, $E_g$ approaches 95%. The intrinsic efficiency of a NaI(Tl) detector is dependent on the size of the crystal—the larger the crystal the higher the intrinsic efficiency of a given energy $\gamma$-ray. A number of well counters are available commercially in various sizes. The so-called standard well counter (1¾″ diameter, 2″ in height with a hole ¾″ in diameter and 1½″ deep) enjoys the most popularity in nuclear medicine. For high-energy $\gamma$-rays (>500 keV), however, one achieves a far better count rate with a 3″ × 3″ well counter than with the standard model. Table 9–1 lists

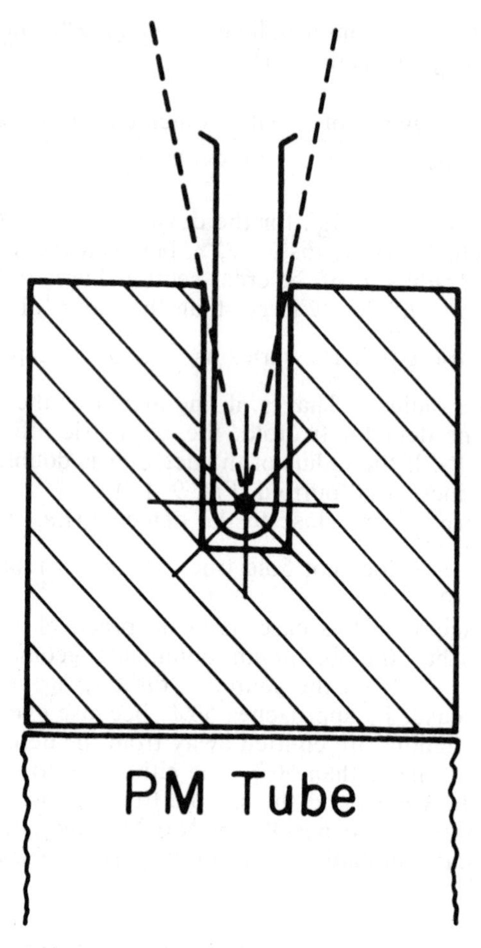


**Fig. 9–2.** Geometric efficiency of a well-type detector. Only a small fraction of radiations that are emitted in the direction enclosed by two dotted lines miss the detector. In such an arrangement, geometric efficiency is close to 95%.

the intrinsic efficiencies of standard and 3″ × 3″ well counters for various γ-ray energies. It can be seen from columns 2 and 3 that at higher energies one gains a factor of 1½ in intrinsic efficiency by using a 3″ × 3″ rather than a standard well counter. For many procedures, one is interested only in the photopeak counts (*i.e.*, one rejects γ-ray interactions via the compton effect). In this case, the advantage of 3″ × 3″ over a standard well counter for higher-energy γ-rays becomes even greater (see columns 4 and 5 in Table 9–1).

The overall efficiency of a well counter can be easily calculated. The geometric efficiency for a small radioactive sample (<1 ml) is approximately 95%; the intrinsic efficiencies for various γ-rays are given in Table 9–1. For a γ-ray with an energy of 140 keV a standard well counter will have an

**Table 9–1. Efficiency of a Well Counter**

| Energy keV | *Intrinsic Efficiency* | | *Photopeak Efficiency* | |
|---|---|---|---|---|
| | *Standard Well* | *3″ × 3″ Well* | *Standard* | *3″ × 3″ Well* |
| 80 | 97* | 98* | 97* | 98* |
| 140 | 94 | 98 | 88 | 96 |
| 280 | 61 | 80 | 49 | 70 |
| 320 | 51 | 73 | 36 | 59 |
| 360 | 48 | 68 | 31 | 50 |
| 410 | 43 | 66 | 24 | 45 |
| 510 | 38 | 59 | 17 | 36 |
| 660 | 32 | 51 | 12 | 25 |
| 880 | 29 | 46 | 8 | 17 |
| 1110 | 28 | 45 | 7 | 16 |
| 1170 | 25 | 42 | 6 | 15 |
| 1270 | 24 | 40 | 5 | 14 |

* Attenuation in sample and sample holder will reduce these numbers slightly.

overall efficiency, E, equal to $0.95 \times 0.94 = 0.89$, *i.e.*, 89%; the overall *photopeak* efficiency will be simply $0.95 \times 0.88 = 0.84$ (84%). One can now also determine the count rate for the photopeak of $^{99m}$Tc (140 keV) for a $\mu$Ci sample. A $\mu$Ci sample of $^{99m}$Tc emits a total of $3.7 \times 10^4 \times 0.88$ $\gamma$-rays per second with an energy of 140 keV; 0.88 is the number of 140 keV $\gamma$-rays per disintegration of $^{99m}$Tc, $n_i$ (see Chapter 2, p. 22). Therefore, the photopeak count rate obtained by a standard well counter will be $3.7 \times 10^4 \times 0.88$ multiplied by the overall photopeak efficiency (84%), or

$$\text{Photopeak count rate} = 3.7 \times 10^4 \times 0.88 \times 0.84 \text{ counts/sec}$$
$$= 2.73 \times 10^4 \text{ counts/sec}$$
$$= 1.6 \times 10^6 \text{ counts/min}$$

In these calculations, we have ignored the absorption effects of the thin aluminum container of the crystal as well as the self-absorption in the solution and the absorption by the walls of the test tube containing the sample. In practice, therefore, one will obtain a count rate slightly less than the one calculated above. However, for the present, assuming the overall efficiency as $1.6 \times 10^6$ counts/min, 10 picocuries ($10^{-5}$ $\mu$Ci) will yield about 16 counts/min. This is a realistic lower limit for the detection of $^{99m}$Tc radioactivity by a well detector and clearly demonstrates the high sensitivity of this device. The actual lower limit of detection, however, will depend on the room background and the availability of the time for a given measurement.

The above discussion about overall efficiency pertains only to small sample volumes (<1 ml). How does the overall efficiency vary with the volume of a sample? This can be ascertained from Fig. 9–3 where the variation in the overall efficiency of a well counter is shown in relationship to the change in the sample volume (keeping the total activity constant). The addition of 1 ml of dilutant to successive samples progressively reduces the geometric

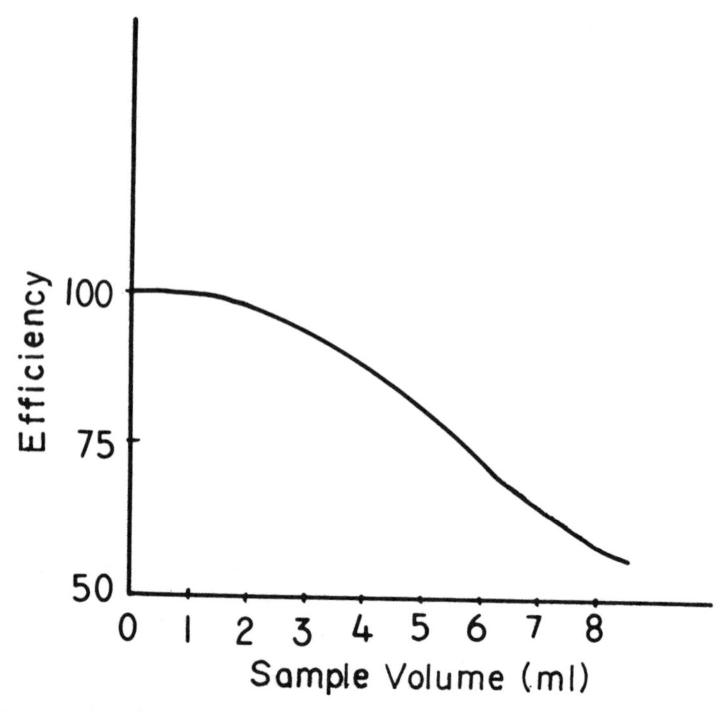


**Fig. 9–3.** Geometric or total efficiency of a well-type detector as a function of the sample volume (keeping the amount of radioactivity constant).

efficiency. Since the intrinsic efficiency remains constant, the overall efficiency behaves in a fashion similar to the geometric efficiency with the increase in sample volume. The overall efficiency for the same amount of radioactivity in a 4-ml volume as compared to a 1-ml volume is about 88% of the latter. Therefore, for a given amount of radioactivity it is preferable to have the volume of a sample smaller than 2 ml.

The other important parameter of a well counter is the dead time, which limits the maximum amount of radioactivity that can be measured without significant (<5%) count loss. Fig. 9–4 shows the relationship of the true count rate to the observed count rate for a typical standard well counter. A good rule of thumb in order to avoid counting loss due to dead time in a well counter is to keep the total (not just the photopeak) count rate below one million counts per minute. This corresponds to roughly 1 $\mu$Ci of radioactivity for the common radionuclides such as $^{131}$I or $^{99m}$Tc. Hence, any sample containing more than 1 $\mu$Ci of radioactivity should not be counted inside the well unless one has taken into account the dead time loss. In that case, a simpler alternative to the correction of dead time loss is either to dilute the sample, using a small aliquot of it for counting, or to use another geometric technique such as counting the sample outside the well.

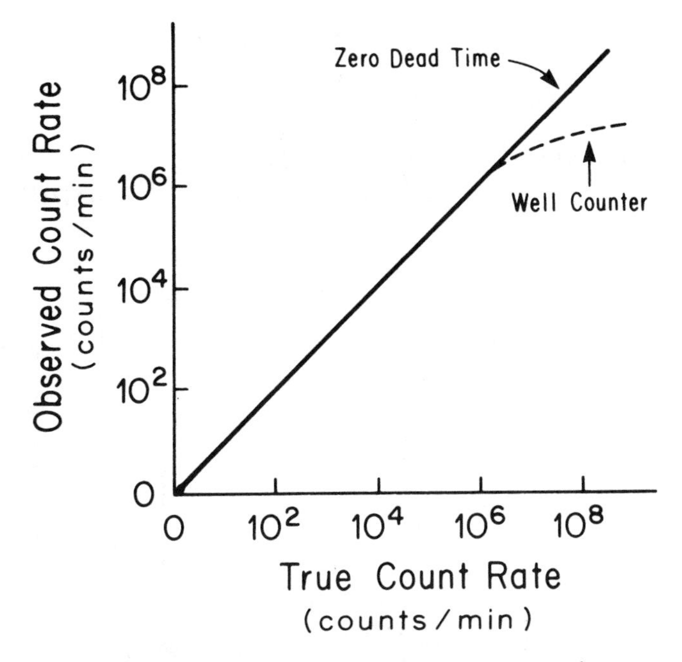


**Fig. 9–4.** Effect of the dead time of the scintillation detector on the count rate.

## LIQUID SCINTILLATION DETECTORS

Six elements—hydrogen, carbon, nitrogen, oxygen, phosphorus and sulfur—make up more than 97% of the total human body weight. There is, therefore, considerable interest in the radioisotopes of these elements both for research and clinical use. The number of radioisotopes of these elements which are easily available and have half-lives long enough for their routine and widespread use is, however, limited to four; $^3H$, $^{14}C$, $^{32}P$, and $^{35}S$. These radionuclides emit only $\beta$-rays ($\beta^-$); they emit no x- or $\gamma$-rays. The detection of charged particles ($\beta$ particles, conversion electrons, or $\alpha$ particles) is much more complicated than the detection of x- or $\gamma$-rays since charged particles have short ranges in solids and liquids, leading to their absorption in the radioactive sample itself and in the walls or window of the detector before detection. To avoid this problem, it is necessary either to prepare an extremely thin sample and use a detector with a very thin window, or to somehow mix the detector and the radioactive source together so that the absorption problem does not arise. This latter alternative is the basis of liquid scintillation counting of $\beta$ or other charged particles. Liquid scintillation counting, because of its ease and versatility compared to other methods, is the preferred method for $\beta$ particle detection, in particular for $^3H$ and $^{14}C$ which emit relatively low-energy $\beta$ particles.

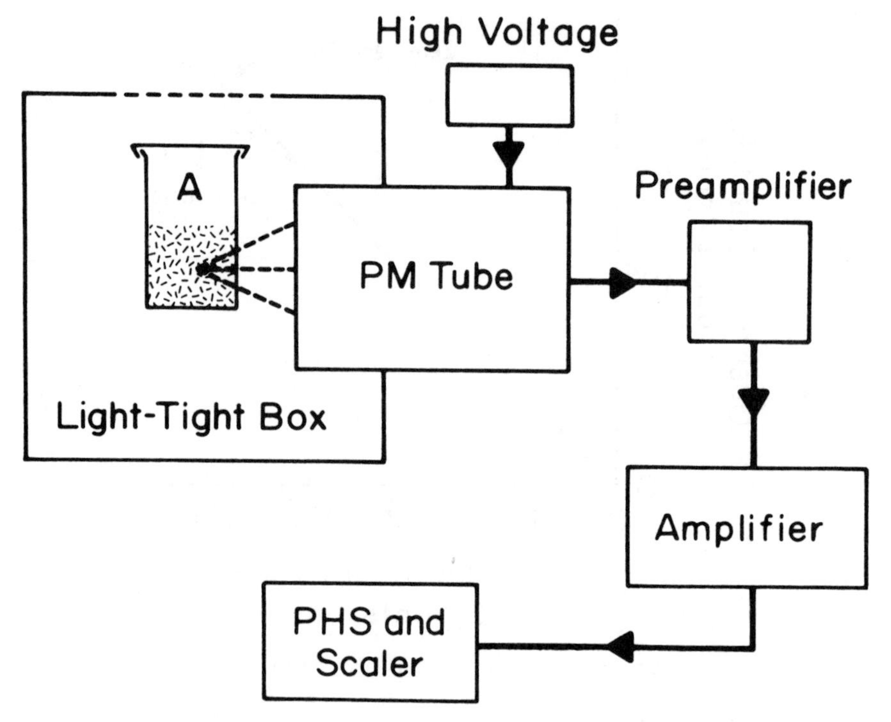


**Fig. 9–5.** Schematic presentation of a liquid scintillation detector. The electronics used in this case are identical to those used in a NaI(Tl) scintillation detector. However, in a liquid scintillation counter, the radioactive sample and the scintillator are mixed together with the help of a solvent in a sample detector vial (A).

## Basic Components

A liquid scintillation detector (see Fig. 9–5) consists of two basic parts: (1) a sample detector vial, and (2) photomultiplier (PM) tube and its associated electronics (see also in Fig. 8–5, p. 111). Since the latter was discussed in detail in Chapter 8, we shall restrict ourselves here to a description of the sample detector vial. This vial consists of the radioactive sample and a suitable scintillator dissolved in a common solvent to form a solution as colorless as possible. The scintillator molecules in this solution act as a radiation detector. Homogeneous mixing of the radioactive sample and detector in this manner has two advantages: (1) since each radio atom of the sample is practically surrounded from all directions by scintillator molecules, the geometric efficiency of such a device is close to 100%; and (2) there is little material between the radiation source and the detector (scintillator) to cause the loss of some $\beta$ particles except for the presence of solvent molecules, which in this case aids in the transfer of energy to the scintillator molecules. Since most effective scintillators are solids, a solvent is essential to achieve a homogeneous mixture of the scintillator and radioactive sample.

The interaction of $\beta$ particles in the solution with the scintillator, as well as with solvent molecules, results in the loss of energy, some of which is then converted to light by the scintillator molecules. The amount of light produced in this case, as in an NaI(Tl) scintillator, is directly proportional to the amount of energy lost. Since a $\beta$-ray has a short range in liquids and loses all its energy in the solution, the amount of light produced is proportional to the energy of the $\beta$-ray. The voltage of the pulse produced by a PM tube in turn is directly related to the amount of light incident on the PM tube. Consequently, the voltage of the pulse produced is directly proportional to the energy of the $\beta$-ray. Pulse-height analysis of the pulses produced then allows the simultaneous use of two or more radionuclides with differing $\beta$-energy spectra in the same sample (for example, $^{3}H$ and $^{14}C$).

The sample detector vial and the PM tube are enclosed in a light-tight compartment to exclude the room and other stray light from reaching the PM tube. In modern liquid scintillation counters the sample detector vial is viewed by two PM tubes rather than one. In this arrangement, using coincidence circuits, electronic noise can be reduced significantly, thus enhancing the sensitivity of low-energy $\beta$ particles which otherwise will be lost in the electronic noise.

### Preparation of Sample Detector Vial

The main problem in the use of a liquid scintillation detector lies in the proper preparation of the sample detector vial. This requires a careful selection of the scintillator and the solvent.

***Selection of Scintillator:*** A good scintillator should have a high light-conversion efficiency, be sufficiently soluble in the solvent of choice, and be chemically stable under a variety of environmental conditions (temperature, humidity, lighting). Among the common scintillators used in liquid scintillation counting, 2,5-diphenyloxazole (PPO), a p-terphenyl and 2,5-bis-2(5-t-butylbenzoxazolyl)-thiophene (BBOT) enjoy the most popularity. Usually a small amount of another chemical known as a secondary scintillator is added to the primary scintillator in the solution. The purpose of a secondary scintillator is to absorb light photons emitted by the primary scintillator in the short-wave length regions (UV and violet) and to re-emit them at longer wave lengths (blue, green or yellow) which can then be more efficiently detected by a PM tube. The compound 1,4-bis-2(5-phenyloxazolyl-benzene (POPOP) is widely used as a secondary scintillator.

***Selection of Solvent:*** The choice of a solvent is dictated by the following requirements: (1) energy deposited in the solvent must be efficiently transferred to the scintillator molecules; (2) the solvent must be transparent to the light produced by the scintillator; and (3) the solvent should be able to dissolve a variety of compounds and be usable at a wide range of temperatures. Toluene, xylene and dioxane fulfill these requirements and are therefore widely used as solvents.

### Problems Arising in Sample Preparation

The problems connected with liquid scintillation counting do not end with the proper selection of a scintillator and a solvent. The sample, generally a biological tissue, has to be processed in such a way that an almost colorless solution is produced when mixed with the scintillator and solvent. This is achieved either by digesting the tissue in a solubilizing agent such as hyamine (a quaternary amine) or by the more involved technique of combustion and oxidation to produce $^{14}CO_2$, $^{3}H_2O$ and $^{35}SO_2$ which are then easily absorbed or dissolved in a suitable scintillator solvent system. Although the hyamine method has the advantage of relative ease over the combustion and oxidation method, it entails additional problems due to a phenomenon known as "quenching."

***Quenching:*** Quenching can be described as any process which interferes with the production or transmission of light from the sample detector vial. This can happen either by chemical quenching or color quenching.

In chemical quenching the presence of trace amounts of certain chemicals (generally present in the tissue or biological sample) interferes with the transfer of energy from $\beta$-rays to the scintillator molecules. Transfer of energy from $\beta$-rays to the scintillator is mediated by short-lived chemical species known as radicals. Chemicals causing chemical quenching also react with these radicals, but in a non-light-producing manner, thereby reducing the amount of light produced. In color quenching the presence of various colored substances in the sample vial reduces by absorption the amount of light produced, hence lowering the amount of light transmitted to the PM tube.

The end result of this loss of light due to quenching is the diminution in the overall efficiency of the system. This diminution in efficiency varies from one sample to another, depending upon the degree of quenching present. For quantitative work one has to take into account this variation in the overall efficiency. This can be done by counting the various samples first and then recounting them after the addition of a known amount of a standard radioactive substance. The effect of quenching can also be corrected by the use of an external $\gamma$-ray source and what is known as the channel ratio method. The details of this method are out of the scope of this book.

***Photo- and Chemiluminescence:*** Two other effects which may interfere with the efficiency of liquid scintillation detectors are photoluminescence and chemiluminescence. Because of the property of certain molecules which lets them absorb light and re-emit it at a later time, the sample detector vial continues to emit light for a short time even when it has been placed in a dark room. This phenomenon is called photoluminescence. "Dark adapting" the samples for several hours before counting generally eliminates this problem.

In chemiluminescence, when two chemicals are mixed or involved in a reaction some light may be produced, depending on the nature of two chemicals involved. Many biological samples, when mixed with toluene or other solvents, produce chemiluminescence. The only way to avoid the production of light through this effect is to wait for the chemical reaction to be completed (which may take up to several days) before counting.

## PROBLEMS

1. How can one increase the geometric efficiency of a detector? Does it depend on the type of the detector being used? What are the advantages of $4\pi$ geometry?

2. Why do the decay rate of a sample and the observed count rate generally differ? Which one of the two is greater generally?
3. Will the thickness of a sample affect the determination of its radioactivity?
4. Why is the determination of radioactivity of a pure $\beta$ emitter sample more problematic than a $\gamma$ emitter?
5. What is the geometric efficiency in a liquid scintillation detector?

# 10

# In-Vivo Radiation Detection —Part I—Basic Problems, Probes and Rectilinear Scanners

In-vivo detection of radioactivity using external detectors constitutes a major concern of nuclear medicine. This involves a variety of studies which can be divided into two subgroups, organ uptake and organ imaging. In organ uptake studies we are interested in the uptake of radioactivity by an organ as a whole, either at a given time (static) or as a function of time (dynamic). Some examples of organ uptake studies are radioiodine uptake by the thyroid, renograms, cardiac output measurements, and blood flow determinations. In organ imaging studies we are interested, not in the uptake of radioactivity by an organ as a whole, but in the relative distribution of radioactivity in the organ either at a given time (static) or as a function of time (dynamic). For example, in liver imaging one is interested in the distribution of radiolabeled colloid in various parts of the liver rather than the total uptake of colloid by the liver. Other examples of this group are bone, brain, lung, spleen, kidney and thyroid imaging.

Even though the two types of studies (uptake and imaging) have quite different aims, some of the problems of in-vivo detection are common to both, and the same radiation detector, a NaI(Tl) scintillation detector is used in both. Therefore, in this chapter, I first discuss the common problems encountered in in-vivo detection. Then I briefly touch upon the specialized instruments (probes) for organ uptake measurement and follow with a brief description of a rectilinear scanner, which is used for organ imaging. Both of these instruments are rarely used in practice but form an important link in the historical development of nuclear medicine instrumentation.

## BASIC PROBLEMS

The use of external detectors for in-vivo measurement of radioactivity automatically excludes the use of radionuclides which do not emit penetrating radiation. Therefore, radionuclides which emit x- or $\gamma$-radiation must be used. Three types of problems arise in the in-vivo detection of such radionuclides: collimation, scattering, and attenuation. All the three factors, when coupled to the fact that the detector has to be a certain distance away from the radioactive source (the source being in-vivo), reduce the geometric efficiency in in-vivo detection by two to three orders of magnitude as compared to the in-vitro case (Chapter 9). This is one of the reasons why millicurie (MBq) amounts are used in in-vivo studies as compared to only microcurie or less (KBq) in in-vitro studies.

***Collimation:*** Since one is interested in the detection of radioactivity present in a small area or volume (*e.g.*, an organ or part of an organ), it is important to exclude all the x- or $\gamma$-rays originating outside the area or volume of interest from reaching the radiation detector. This is achieved by the use of a collimator, a device that limits the field of view of a radiation detector. A variety of collimators are available for different purposes. These are usually made of lead, which is inexpensive and possesses both a high density and a high attenuation coefficient for the x- or $\gamma$-rays whose energies are of particular interest in nuclear medicine ($<$500 keV). A diagram of a simple collimator is shown in Fig. 10–1. Note that simple collimation does not allow for full discrimination against the radioactivity underlying or overlying the volume of interest. Also, the field of view of such a collimator is determined by two parameters, the length and the radius of the opening

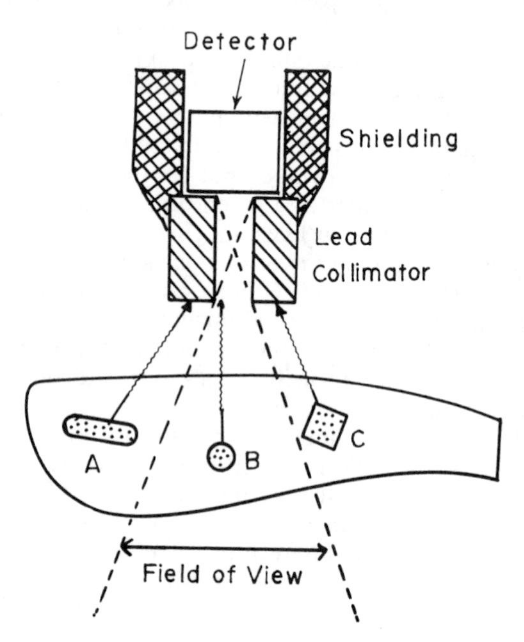


**Fig. 10–1.** A simple collimator. The collimator restricts the field of view of a radiation detector. Only those $\gamma$-rays which originate from the radioactive source B can reach the detector, whereas all $\gamma$-rays originating from radioactive sources A and C are blocked by the collimator. The field of view of this type of collimator increases with distance from the collimator, as shown by the dotted lines.

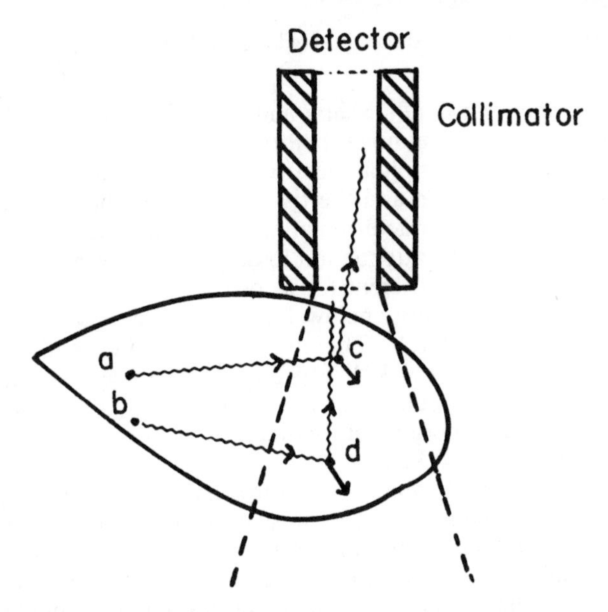


**Fig. 10–2.** Compton scattering of γ-rays interferes with the function of a collimator. Gamma-rays originating outside the field of view from points *a* and *b* are able to reach the detector as a result of compton scattering at points c and *d*, respectively. The only effective way to reject such events is by the use of pulse-height analysis. The solid arrows at the points c and *d* represent the compton scattered electrons. The dotted lines show the field of view of the collimator.

(hole) in a collimator. By reducing the radius or increasing the length, one can reduce or narrow the field of view of a collimator to any desired size. The field of view of a single hole is related to the resolution and sensitivity of a collimator.

***Scattering:*** The γ- and x-rays emitted by radionuclidic sources embedded in a mass of matter experience, during interaction with that material, scattering via the compton process. In compton scattering (see Chapter 6), an interacting γ- or x-ray loses some of its energy and changes its direction. The change in the direction of many γ-rays which originate outside the field of view of the collimator, causes them to be scattered toward the radiation detector, thus defeating the purpose of this device (Fig. 10–2). Since, in each scattering event, the x- or γ-ray loses energy, compton-scattered x- or γ-rays can be excluded by employing energy discrimination in such a manner that only those rays which have experienced no energy loss are detected.

When using a NaI(Tl) detector, this is accomplished by having a PHS select only those pulses whose height corresponds to the photopeak of the unscattered γ-ray. To do an effective job of energy discrimination, the energy resolution of a detector should be very good. A NaI(Tl) scintillation detector possesses a moderate energy discrimination capability (FWHM 50 keV for a 660-keV γ-ray) as compared to a Ge(Li) semiconductor detector (FWHM 3 to 5 keV). In spite of their better energy discrimination capability, Ge(Li) detectors are seldom used in nuclear medicine, primarily because of their very low sensitivity compared to a NaI(Tl) scintillation detector (see Chapter 8).

***Attenuation:*** Since the depth, shape and size of an organ containing the radioactive substance are unknown beforehand, the attenuation of x- or γ-rays (absorption through the photoelectric effect and scattering through the compton effect) in the organ and the tissue overlying the organ is another serious hurdle in the in-vivo determination of radioactivity.

It is possible to minimize attenuation loss by the use of high-energy x- or γ-rays. The attenuation coefficient for x- or γ-rays in tissue drops sharply with an increase in γ-ray energy below 100 keV and levels off with increase in γ-ray energy above 100 keV. Therefore, radionuclides emitting x- or γ-rays with energies above 100 keV are preferred. Since the sensitivity of NaI(Tl) detectors and the collimators used in scanning decreases with the increase in x- or γ-ray energy, the optimum range of energies for in-vivo use is between 100 and 300 keV.

In organ uptake studies, in addition to the use of high-energy γ-rays, attenuation effects are taken care of by measuring a known amount of radioactivity in a standard phantom which reflects the average size, shape and depth of an organ in a standard man. In organ scanning studies, however, attenuation is considered a *fait accompli* and is taken into account only when interpreting a scan.

## ORGAN UPTAKE PROBES

An uptake probe consists of two basic parts: a NaI(Tl) detector and a collimator.

***NaI(Tl) Detector:*** In in-vivo studies, the size of the crystal in a NaI(Tl) detector is an important consideration. Crystal size is determined by the energy of the γ-ray to be detected and the sensitivity requirements of a particular study. For thyroid uptake studies using $^{131}I$, the International Atomic Energy Agency recommends that a crystal less than 1″ × 1″ should not be used. A 1½″ × 1″ crystal is generally adequate for thyroid uptake measurements of $^{131}I$ and also serves as a multipurpose instrument in the nuclear-medicine laboratory.

***Collimator:*** The design of a collimator for uptake studies is dictated by its intended application. However, the following general requirements apply in most cases:

(1). In order to keep the radiation burden to the patient to a minimum, the overall efficiency should be as high as possible.

(2). The field of view of the collimator should be well defined but flexible enough to take into account the varying size of a particular organ in different patients while at the same time excluding any radioactivity present in other organs.

(3). Since the distribution of radioactivity within the organ, as well as its size, shape and depth, is not known, the overall efficiency or sensitivity should be uniform across the field of view of the collimator and throughout the thickness of the organ.

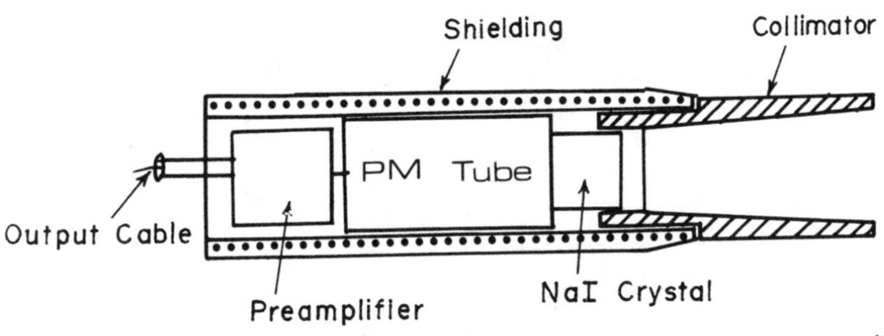


**Fig. 10–3.** A typical NaI(Tl) crystal collimator assembly (probe) used for thyroid uptake of $^{131}$I.

Since requirements 1 and 3 oppose each other to some extent, one makes the best one can of a given situation. A typical collimator used in thyroid uptake measurement is shown in Fig. 10–3. The overall sensitivity within the field of view of such a collimator varies inversely as the square of the distance between the source and the detector, but the uniformity across the organ improves as the distance from the detector is increased. For thyroid uptake, a distance of 30 cm is considered optimum.

## ORGAN IMAGING DEVICES

Administration of a particular radiopharmaceutical results in its selective localization in the organ or organs of interest in a patient. The distribution of the radiopharmaceutical in an organ may vary within the organ itself, particularly as a result of some focal disease. A diseased or abnormal area in the organ may be hotter (more radioactivity) or colder (less radioactivity) compared to adjacent normal tissue. The purpose of organ scanning or imaging is to unravel this relative distribution of the radioactivity present in an organ. Ideally, we wish to delineate this distribution in all three dimensions (volume), but because of various technical problems this is not routinely feasible. Instead, a two-dimensional or areal distribution of radioactivity, which is very useful clinically, is obtained. The lack of three-dimensional information in this case is somewhat compensated for by determining the area distribution from multiple directions, generally four: anterior, posterior, right and left laterals. A two-dimensional record of the distribution of the radioactivity present in an organ is called a scan or an image.

A scanner or imager consists of four elements: a collimator, radiation detector, device to give the location (*i.e.*, x, y coordinates) of the radioactivity, and a system to display the relative distribution in a manner easy to comprehend. Depending on how the x, y coordinates, or information about the location of the radioactivity, are obtained, imaging devices can be grouped into two classes: Rectilinear Scanners and Gamma Cameras (Chapter 11). Here I discuss only the rectilinear scanner. The Gamma Camera is discussed in the next chapter.

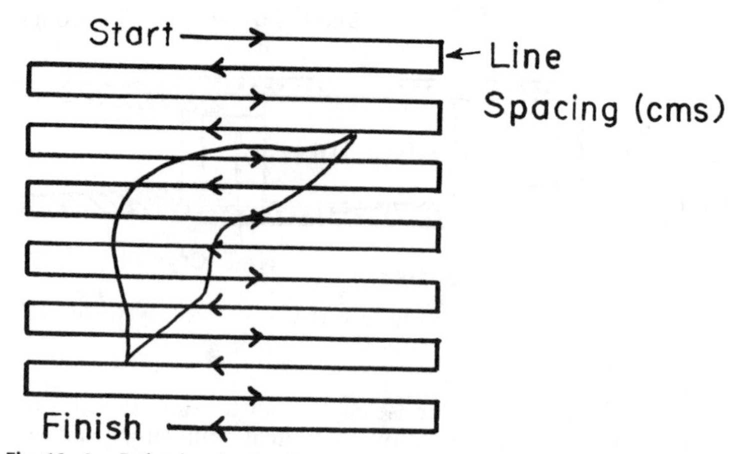


**Fig. 10–4.** Path taken by the detector head in rectilinear scanning of an organ.

## Rectilinear Scanner

In a rectilinear scanner a NaI(Tl) scintillation detector with an appropriate collimator (detector head) is moved over the organ of interest in a straight line with the help of an electric motor. After traversal of a specified distance, the detector head either steps up or steps down a small distance and then continues the straight-line motion in the opposite direction. This back-and-forth motion of the detector head (Fig. 10–4) continues until the area occupied by the organ has been fully scanned. The field of view of the collimator in this case is narrow, and the rectilinear motion of the detector head over the organ generates the span point by point. This type of image formation is similar to that encountered in a television where also the picture is constructed point by point, even though our eyes do not realize it because of the rapidity of the whole process.

The detector head assembly consists of a collimator, NaI(Tl) crystal, and PM tube with preamplifier, all appropriately shielded to reduce room background. Two crystal sizes, 3″ × 2″ and 5″ × 2″, are commonly employed. The collimator used in a rectilinear scanner is known as a multihole focusing type collimator and is shown in Fig. 10–5. It consists of a lead cylinder with a number of tapered holes formed to converge at a single point outside the collimator known as the focus. The side of the collimator with holes of larger radii faces the NaI(Tl) detector, whereas the side with the holes of smaller radii faces the patient. The distance between the collimator face and the focus is the focal length of the collimator.

The information gathered by the detector head regarding the amount of radioactivity at a given location (number of counts per unit of time) is continuously relayed to a photographic system employing a 14″ × 17″ x-ray film, where a small cathode-ray tube, with a fast phosphor screen and well-collimated light spot which exposes the x-ray film, is moved in a light-tight box in synchronization with the detector head. At the completion of the scan, the x-ray film is developed. The count rate (R) observed by the detector

In-Vivo Radiation Detection—Part I—Probes and Rectilinear Scanners

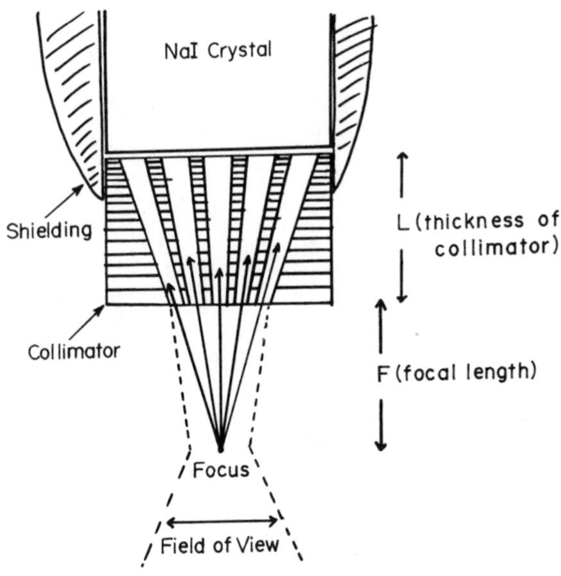


**Fig. 10–5.** Focusing-type collimator. Gamma-rays can reach the detector only through the narrow channels as shown by the solid arrows. The field of view of such a collimator (dotted lines) is in general narrow and is narrowest (about 1 cm) in the focal plane. It becomes wider above or below the focal plane.

head in a given location of an organ is displayed as the darkening of the x-ray film at a corresponding location on the film. Dark areas on the film represent high areas of radioactivity in the organ, whereas light areas on the film denote a low level of radioactivity.

## PROBLEMS

1. List the problems that make quantitation of in vivo radioactivity difficult.
2. How does pulse height analysis help in the detection of in vivo radioactivity?
3. What is the primary function of a collimator? Why are collimators made of lead? Can other materials be used instead of lead?
4. Why is there an optimum range of distance between the patient's thyroid and the probe when measuring the radio-iodine uptake?
5. Why rectilinear are scanners no longer in clinical use?

# 11

# In-Vivo Radiation Detection—Part II—Gamma Camera

A gamma camera—in particular a scintillation camera—occupies a central place in every nuclear medicine department. In a gamma camera, as opposed to a rectilinear scanner where an organ is scanned point by point, the whole organ or a large part of the body is scanned, or more appropriately, imaged simultaneously. In this respect a gamma camera behaves similarly to a photographic camera, even though the two types of cameras are entirely different in construction and operation. Unlike light rays, x- or $\gamma$-rays can not be reflected or refracted by employing mirrors, lenses or prisms. Therefore, the general principles of light photography can not be applied to imaging of objects emitting x- or $\gamma$-rays. Instead, the selective absorption and transmission of x- and $\gamma$-rays by different materials, such as lead and air, forms the basis of imaging with a gamma camera.

The simultaneous visualization of an entire organ or organs by a gamma camera is its single most important quality and has made rectilinear scanners obsolete. This feature makes the study of rapid dynamic processes possible. Dynamic studies with 10 to 20 images per second are now routinely obtained to determine cardiac output and ejection fraction.

Of a variety of approaches attempted in research laboratories for the development of gamma cameras, the scintillation camera developed by Anger has emerged as a superior choice in clinical nuclear medicine.

## SCINTILLATION CAMERA

In a scintillation camera, a large, disc shaped, single NaI(Tl) crystal is viewed from one side by an array of PM tubes. Such an array of PM tubes not only determines the total amount of light produced by a gamma ray interaction, but also the location of light production in the crystal.

Physically, a scintillation camera is divided in two parts: (1) the detector head containing the collimator and the NaI(Tl) crystal with PM tubes (and associated electronics) mounted on a stand where it can be easily moved up and down or rotated in any desired position with the help of hand-held controls; (2) the console, which houses the power supplies and operational controls of the scintillation camera including the display module. In some scintillation cameras called portable scintillation cameras, the detector head and console are joined together in one unit so that it can be moved from one location to another location without too much difficulty. Operationally, a scintillation camera consists of four basic parts: (A) Collimator, (B) Detector, (C) X, Y or position determining circuit and (D) Display. Here I describe the workings of a scintillation camera only. Its operating characteristics such as resolution,* sensitivity, and uniformity and quality control are discussed in Chapter 12.

### Collimators

The purpose of a collimator in an Anger camera is to allow x- or $\gamma$-rays originating from a selected area of an organ to reach a selected area of the detector. Thus a collimator establishes a one to one correspondence between different locations on the detector and those within the organ. Another feature of a scintillation camera collimator is that its field of view is large enough to encompass completely the total organ or the desired part of the body to be imaged. Four types of collimators have been employed with Anger Cameras: pinhole, parallel hole, and converging and diverging (Fig. 11–1). As can be seen, in all these collimators, gamma rays originating from one area of the arrow reach only a selected area on the crystal. Thus, gamma rays originating from the front of the arrow reach a different location on the crystal than gamma rays originating from the middle or back of the arrow. The size of the image formed on the crystal depends upon the type of collimator and distance of the object (arrow) from the collimator.

The choice of a particular type of collimator is basically dictated by the size of the organ to be imaged. For imaging organs which are similar in size to the size of detector (NaI(Tl) crystal), parallel hole collimators provide the best sensitivity and resolution. For organs larger than the size of the crystal, diverging collimators are preferred. For organs smaller than the size of the crystal, converging collimators have shown great merit. When the size of the organ is small, a pinhole collimator may be equally satisfactory.

One problem which makes the use of pinhole, converging or diverging collimators less satisfactory compared to parallel hole collimators is the fact that for three dimensional objects (which all organs are), the different planes of the object (front, back or middle of the organ) are magnified or minified to different degrees by these collimators. This produces distortions in the image which, under many circumstances, are unacceptable.

Commercially, the collimators, besides being characterized by the above four types, are also classified according to their resolution or sensitivity and the energies of gamma rays for which they have been optimized. The resolution and sensitivity of a parallel-hole collimator and the factors which affects them are discussed in Chapter 12.

---

* Wherever the word "resolution" appears in this chapter, it should be read as "spatial resolution."

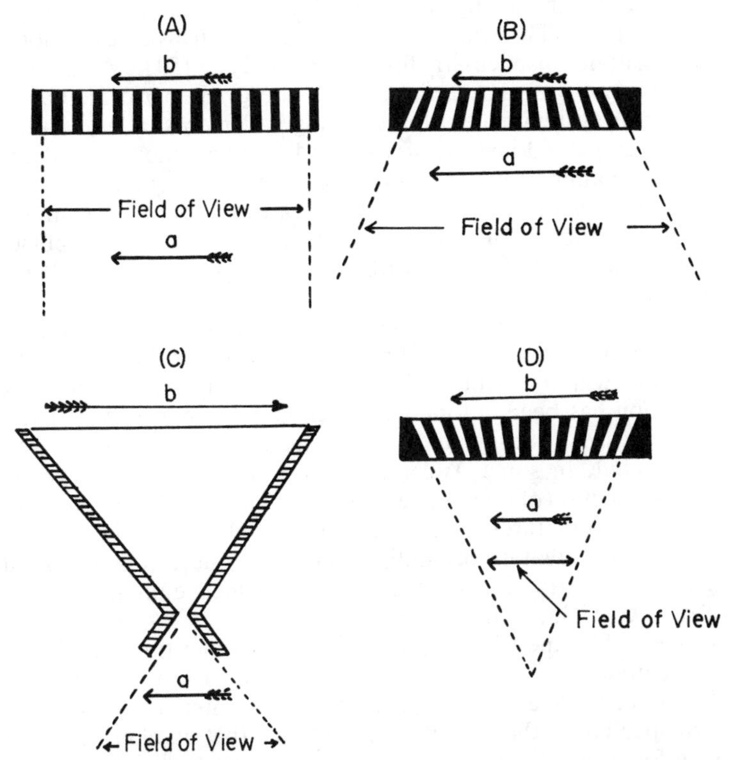


**Fig. 11–1.** Collimators used in a scintillation camera. (A) *Parallel-hole collimator:* The object, *a*, projects the same size image, *b*, on the crystal face. The field of view of such a collimator does not vary significantly with distance from the collimator. (B) *Diverging collimator:* The size of the image, *b*, is smaller than the size of the object, *a*, and the field of view increases as one moves away from the collimator. (C) *Pinhole collimator:* A magnified or minified image, *b*, of an object, *a*, is produced, depending on its distance from the pinhole. The field of view of a pinhole increases rapidly as one moves away from the pinhole. (D) *Converging collimator:* This collimator produces a magnified image, *b*, of an object, *a*. The field of view decreases as one moves away from the detector. A converging collimator provides the optimum sensitivity and resolution for an object which is smaller than the crystal size used in the scintillation camera.

## Detector, NaI(Tl) Crystal

As has already been pointed out, the basic detector element in an Anger Camera is a large, disc shaped [NaI(Tl)] crystal which is viewed from one side by a large number of PM tubes. The diameter of the crystal varies from 11″ to 20″. Eleven inch diameter crystals are employed in the standard Anger Camera, whereas 16″ to 20″ diameter crystals are employed in so called large field of view (LFOV) Anger Cameras. The main advantage of a large crystal is increased sensitivity for large organs such as lungs or the whole body. For more effective use of the crystal area, rectangular crystals are also available in some scintillation cameras. The thickness of the crystal is generally ½″ but recently Anger Cameras with ⅜″ or ¼″ thick crystals have been introduced, particularly for nuclear cardiology work. The reduced thickness of the crystal results in the improvement of intrinsic resolution (to be defined in the next section). The trade off for improved intrinsic resolution is reduction in sensitivity, particularly for higher energy (>150 keV) gamma rays.

The detection of the gamma ray and its energy, is performed as in any NaI(Tl) detector system except that, in Anger Cameras, a large number of PM tubes are employed instead of a single PM tube. In order to determine the energy of a gamma ray, one has to determine the total amount of light produced in the crystal. In an Anger Camera this is done by summing the output of all PM tubes. The summated pulses are known as Z pulses and pulse height analysis on Z pulses with a pulse height selector (PHS) allows us to select the pulses which have the desired energy. Thus, like any other NaI(Tl) detector, in an Anger Camera also there are four controls related to the detection of gamma rays: High voltage, Gain of the amplifier, Peak energy, E, and window width $\Delta E$ or %E. In many Anger Cameras, the energy selection has been automated; to choose a gamma ray of particular energy, one presses a designated button and appropriate pulses are automatically selected. In some Anger Cameras, there is also a provision for simultaneous selection of two or even three gamma rays of different energies. These employ two or three PHS's, instead of only one as in the standard Anger Camera. This feature of multiple gamma ray detection is quite useful for imaging the distribution of radionuclides which emit more than one gamma ray, e.g., $^{67}$Ga, $^{111}$In or even $^{201}$Tl.

Pulses selected by the PHS are then fed into a scaler timer module which allows the scintillation camera either to operate for a fixed time interval (preselected time) or to detect an assigned number of gamma rays (pre-selected counts) irrespective of the time it takes to reach that number. There is also a provision for Anger Cameras to stop at preselected time intervals or preselected number of counts, depending on which happens first. A manual control allows one to start or stop counting at any time one wishes. Another feature found only in some scintillation cameras is pre-selection of the information density in a given area of the image. The scintillation camera will stop when pre-selected number of counts have been acquired in the desired area of the image.

## Position Determining (X, Y) Circuit

A collimator in an Anger Camera allows $\gamma$- or x-rays originating from one small part of an organ to reach a small part of the crystal in a one to one correspondence. To keep this correspondence intact, we should know where the gamma rays are interacting in the crystal. This generally is accomplished with the help of a large number of PM tubes, but in Fig. 11–2, it is illustrated by considering a simple array of five PM tubes. In this case, when light is produced at a point *a* in the crystal, PM tube 1, closest to the point *a*, receives the maximum amount of light. Similarly when light is produced at points *b* and *c*, PM tubes 2 and 3 respectively, will receive the largest amount of light. Thus, by knowing which PM tube received the largest amount of light, it is possible to know the rough location (near the PM tube receiving the greatest amount of light) of the point of light production. Now, if we wish to locate the point of light production more accurately, the amount of light

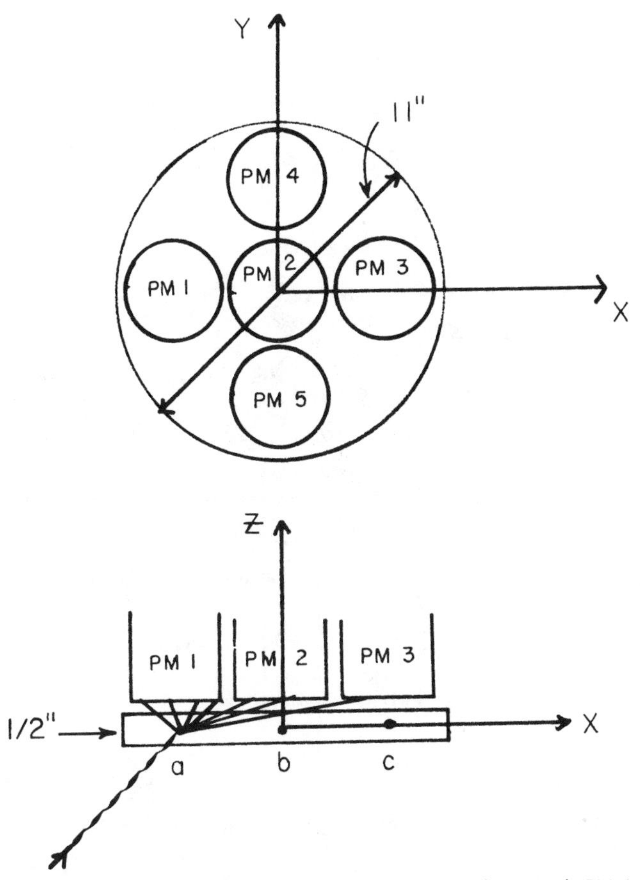


**Fig. 11–2.** Anger camera. When a γ-ray interacts at point *a* in the crystal, PM 1 receives the maximum amount of light, thus providing the approximate location of the point at which the light is produced. Similarly, when interaction takes place at points *b* and *c*, PM 2 and PM 3 receive the most amount of light, respectively. If the amount of light received by each PM tube is known, rather than the PM tube which receives the maximum amount of light, the location at which light is produced in the crystal can be determined more accurately.

received by each PM tube rather than the one receiving the most light (i.e., the distribution of light among different PM tubes) has to be taken into account. The distribution of light among different PM tubes will be directly proportional to the solid angle subtended by each PM tube at the point of light production. This fact is employed in the calculation of the exact location of the point of light production.

To accomplish this in practice, the outputs of various PM tubes are summed with appropriate weighting factors to yield four analog signals, known as $X^+$, $X^-$, $Y^+$ and $Y^-$. This is illustrated in Fig. 11–3 for 5 PM tubes. In commercial Anger Cameras, the number of PM tubes varies from 19 to 96 and the summation circuits are more complex and may differ from

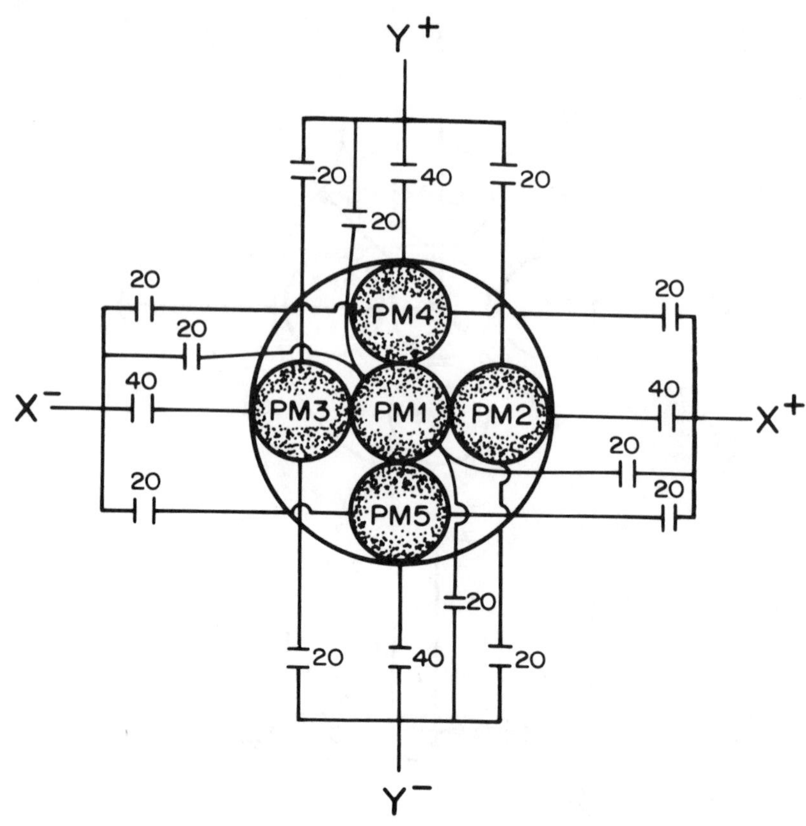


**Fig. 11–3.** A simplified diagram for generating X, Y localizing signals, $X^-$, $X^+$, $Y^-$ and $Y^+$, for a 5 PM tube Anger camera.

one manufacturer to another. However, in all of them, the position defining voltages X and Y, as well as the Z pulses (previous section) are generated by the four voltages $X^+$, $X^-$, $Y^+$ and $Y^-$ as follows:

$$Z = X^+ + X^- + Y^+ + Y^- \tag{1}$$

$$X = \frac{K}{Z}(X^+ - X^-) \tag{2}$$

$$Y = \frac{K}{Z}(Y^+ - Y^-) \tag{3}$$

where K is a constant.

The pulses X and Y, which are directly proportional to the distance along the x- and y-axis of the point of light production in the crystal, are used to deflect a light spot on a cathode ray tube or oscilloscope in direct proportion to the amplitude (magnitude) of X and Y. From the oscilloscope, the final image of the distribution is formed, as discussed in the next section.

It should be pointed out here that, even after considering the amount of light received by all the PM tubes, there is always a small error involved in the exact localization of the point of light production. This error is a measure of the intrinsic resolution of the Anger Camera. Intrinsic resolution is a complex function of the thickness of the crystal, number of PM tubes employed for position determination, type and shape of PM tubes, and the thickness of light pipe, if used, to couple the PM tubes with the crystal. Reducing the thickness of the crystal improves the intrinsic resolution, but it also decreases sensitivity of Anger Camera as lesser numbers of gamma rays interact in the crystal. Therefore a compromise has to be made between the intrinsic resolution and sensitivity of the Anger Camera. The optimum range of thickness for scintillation cameras is ⅜″ to ½″ for 140 keV gamma rays. Another factor which also affects intrinsic resolution of the Anger Camera is the energy of gamma rays. This is due to the fact that a higher energy gamma ray produces more light in the crystal (for photoelectric events which we always select) than a lower energy gamma ray. More light enables better localization of the point of $\gamma$-ray interaction, and better localization means better intrinsic resolution for the higher energy gamma rays.

## Display

Gamma rays originating in the field of view of a collimator interact in the crystal at different locations. These interactions in general occur in random fashion. A display device should be able to portray such a randomly generated position (X,Y) information quickly (at least $10^6$ events per minute) and accurately. A cathode ray tube or an oscilloscope of which a cathode ray tube is an integral part, is effectively used for this purpose.

A cathode ray tube (CRT) is an evacuated glass tube consisting of five basic components: an electron gun, a focusing electrode, horizontal deflection plates (X direction), vertical deflection plates (Y direction), and a phosphor screen. These are shown schematically in Fig. 11–4. The electron gun produces a stream of fast electrons. Number of electrons or the intensity of the electron stream can be varied, if desired, by intensity control, I. The focusing electrode allows focusing the electron stream to a narrow circular beam (about 0.1 mm in diameter). When voltage pulses are applied to the horizontal and vertical plates, the electron beam moves in the X and Y directions in direct proportion to the magnitude of the voltage pulses applied at the horizontal and vertical plates respectively. The duration for which the electron beam stays at its new location depends upon the duration of the voltage pulses applied to the horizontal and vertical plates. It is generally less than a microsecond. When there are no voltage pulses applied to the horizontal and vertical plates, the electron beam remains at the center of the phosphor screen. The location of the electron beam on the screen is made visible by the phosphor which emits light at the point where the electron beam strikes it. In this way, when voltage pulses of different magnitudes are applied in succession to the horizontal and vertical plates, the light spot on the CRT screen moves from one place to another but always at a distance which is directly proportional to the magnitude of the applied voltage pulses.

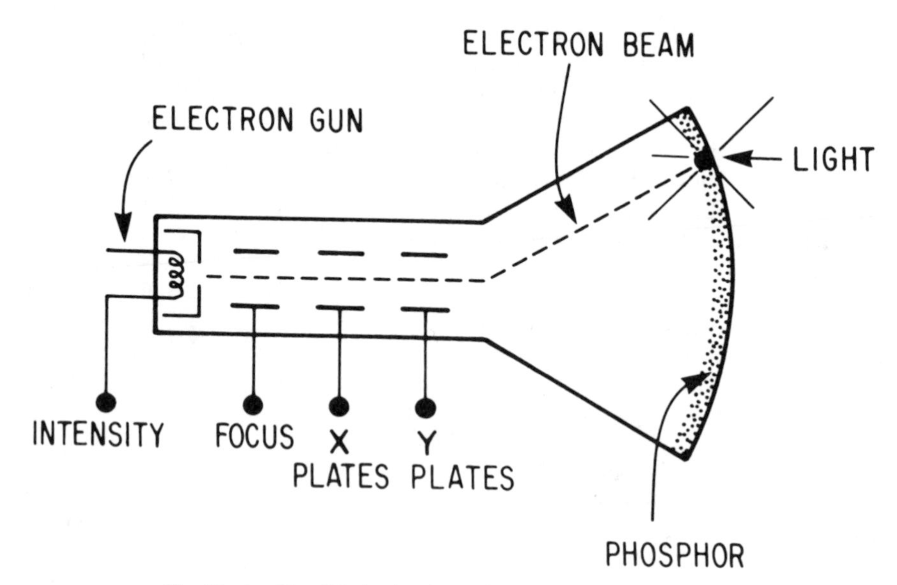


**Fig. 11–4.** Simplified schematic of a cathode ray tube (CRT).

For displaying position information coming from the scintillation detector, the X and Y voltage pulses are applied to the horizontal and vertical plates of a CRT. The Z signal which carries the energy information exercises a veto on the X and Y signals in such a manner that these are applied to the CRT horizontal and vertical plates only if the Z signal is within the energy range selected by the PHS. If the Z pulses are outside the range selected by PHS, then X and Y are not applied to the CRT horizontal and vertical plates. Thus, only those gamma ray interactions which deposit energy in the crystal in the range selected by PHS are displayed on the CRT screen.

In summary, the display functions as follows: A gamma ray interacts in the detector, the detector produces three signals giving the location (X and Y pulses) of the gamma ray interaction and the energy transfer (Z pulse) by the gamma ray interaction. The Z pulse is analyzed and if it is within the selected range, the position signals (X and Y pulses) are applied to the CRT plates which deflect the light spot from the center of the screen to a distance proportional to the X and Y voltages. When a new gamma ray interacts in the crystal, a new set of X, Y, and Z signals is produced which then deflects the light spot to a new location given by these signals. In this way, as more and more gamma rays interact in the crystal, the light spot on the CRT screen keeps moving from one place to another in correspondence with the location of gamma ray interaction in the crystal (Fig. 11–5). Since the size of CRT's usually employed range from 3″ to 5″ in diameter, the image on the CRT screen is displayed in a smaller size than the actual size of the organ or part of the body imaged.

A flying light spot on the screen of a CRT does not constitute an image. This image is formed by point by point integration of this information on a photographic film.

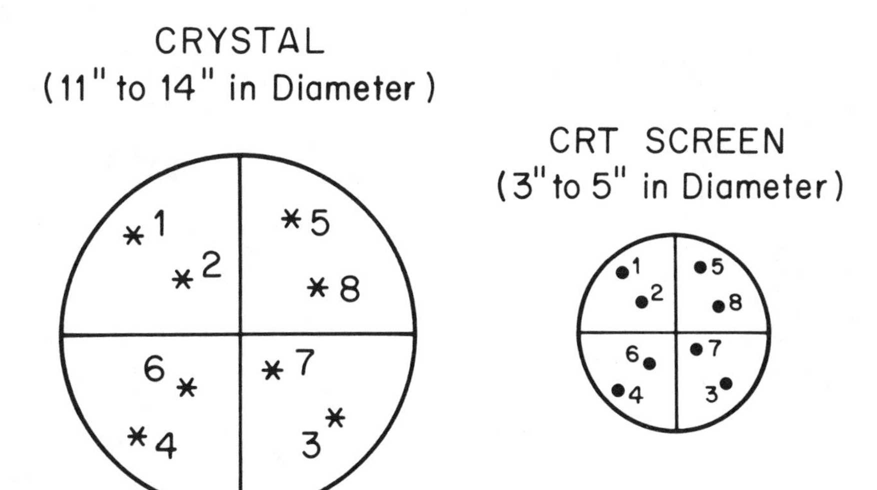


**Fig. 11–5.** Direct correspondence of the location of gamma ray interaction in scintillation detector with the location of light spot on CRT screen.

On a film, darker areas represent more radioactivity, whereas lighter areas represent less activity. Film darkening is quantitatively measured by a parameter known as optical density or, simply, density (D). It is defined as the logarithm (base 10) of the ratio of the intensity of incident light on the film to the intensity of the light transmitted by the film. According to this definition, an area of the film with a density of 2 will transmit only 1% light and will, therefore, appear almost black to the naked eye. A density of 0 represents 100% transmission; therefore, an area with 0 density will appear white. Densities between 0 and 2 will appear as shades of gray. The relationship of D to exposure I for a typical x-ray film is shown in Fig. 11–6, where D is plotted as a function of the logarithm (base 10) of exposure I. This curve is known as the H-D curve of the film. The average slope between points A and B determines the contrast of a film and the relative log exposure between the points A and B (horizontal distance) determines the latitude of a film. A high-contrast film displays smaller-exposure variations than a low-contrast film. But a high-contrast film has smaller latitude. Therefore, the range of exposures that can be displayed on a high-contrast film is smaller (Fig. 11–6).

It can be seen from the H-D curve that the density, D, is dependent on the exposure or the count rate only in the region which lies between points A and B marked on the curves. The trick in photo display is to correspond this region to the range of the count rates which is of the most interest. Normally, count rates in an organ may vary from zero to a maximum $R_{max}$. For effective display, $R_{max}$ (also known as the "hot spot") should correspond to point B, and zero count rate to point A on the H-D curve. The $R_{max}$ for individual patients will differ because of variations in the administered radiopharmaceutical dose, localization and distribution in the organ, and the size and shape of the organ.

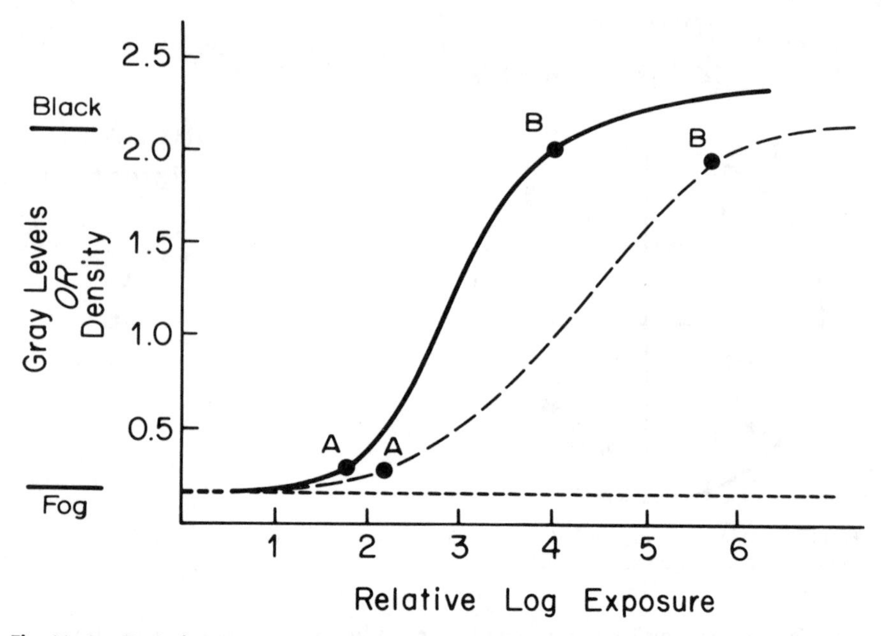


**Fig. 11–6.** Typical H-D curves of x-ray films used to display information. For the proper display of count rate information, the maximum count rate (hot spot) should correspond to point B and the minimum count rate to point A. Two types of films are shown. Solid curve is for a film with high contrast and small latitude. Broken curve is for a film with low contrast and large latitude.

Since there is no convenient way of finding the "hot" spot using a scintillation camera, it is difficult to align the count-rate range of interest in each scan with the A-B portion of the H-D curve. Therefore, particularly for fast dynamic studies, the exposure settings are only a guess which sometimes produces bad results. The best solution to this problem is to interface the scintillation camera with a computer and store images temporarily so that in case of a bad exposure, the image can be recalled and a proper exposure made.

A multiformat recording system is most commonly employed in nuclear medicine. In this system, multiple images are recorded on a single sheet of x-ray film, usually 8 × 10 or 11 × 14 inches in size. The number of images and, therefore, the size of the image which can be recorded on a single sheet can be varied easily with the help of controls provided for such purposes. Thus, a single sheet may contain from 1 to as many as 64 images. The main advantage of this device is that all views from one patient, including dynamic studies, can be recorded on a single sheet of film, thus consolidating most of the information in one place. A typical study, a bone scan using this format, is shown in Fig. 11–7.

## IMAGING WITH A SCINTILLATION CAMERA

Generally, the following steps are taken to obtain an image with a scintillation camera:

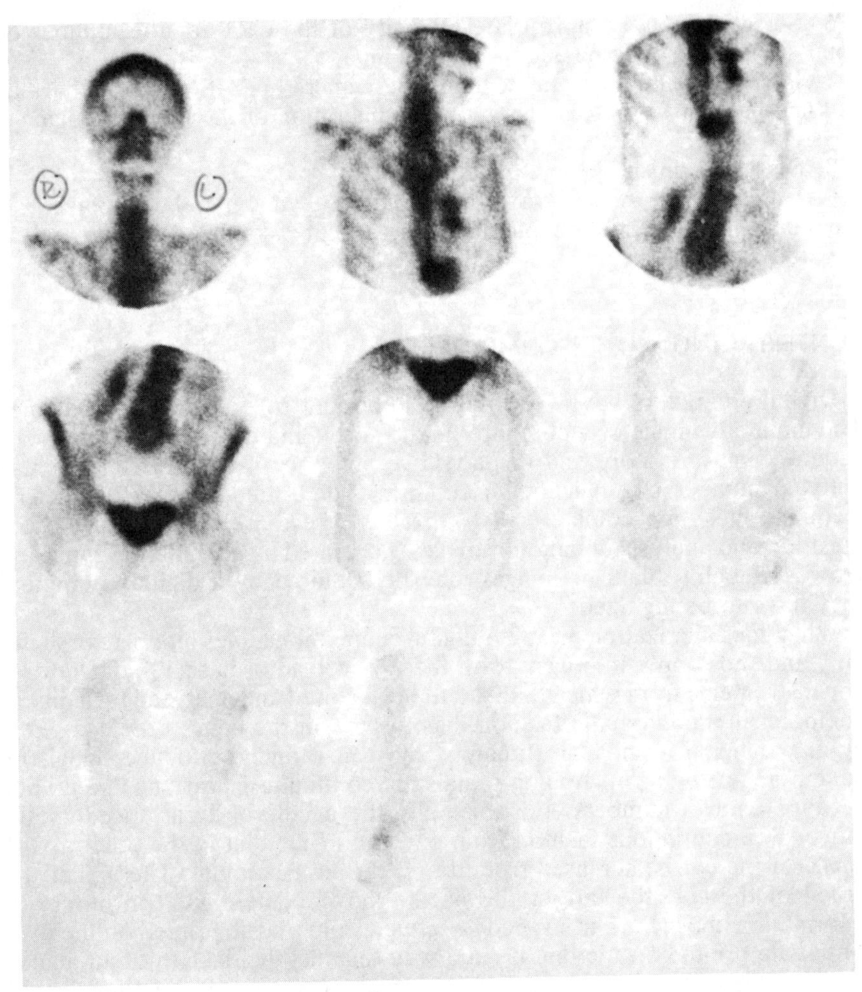


**Fig. 11–7.** Anterior view of a bone scan recorded on a multiformat recording device. In this example only 8, out of possible 9, images were recorded.

1. Selection of the study to be performed, e.g., brain, liver, etc.
2. Selection of the radiopharmaceutical and the dose of radiopharmaceutical. Radiopharmaceutical is generally administered to patients away from the scintillation camera, but sometimes, particularly when fast dynamic studies are to be performed, it may have to be administered with the patient in appropriate position under the camera (step 7).
3. Selection of the PHS parameters (peak energy and % window corresponding to the gamma ray emitted by the radionuclide to be used).
4. Selection of an appropriate collimator (with respect to energy and resolution).
5. Selection of the mode: Accumulation of a certain number of counts or exposure for fixed amount of time.

6. Selection of the appropriate intensity of the CRT for the number of counts expected or to be acquired in the image.

7. Positioning of the patient under the camera and, if the radiopharmaceutical has not been administered, administration of the radiopharmaceutical.

8. Start and finish of the exposure.

9. Development of the film. If more than one view is to be displayed on the same film, then the development of the film takes place at the end of the study.

## INTERFACING WITH A COMPUTER

Digital computers have acquired an important role in nuclear medicine. The main advantage of a computer is the speed and ease with which it can acquire, analyze, store and display large amounts of complex data. Presently, a number of small digital computers dedicated entirely to nuclear medicine tasks are commercially available. These are mostly used in acquisition and analysis of data (or images) generated by scintillation cameras. However, before data or images may be acquired by a digital computer, they have to be digitized.

What does digitization mean? A digital computer handles numbers or digits only and those only in binary form (base 2 instead of base 10). Therefore, any instrument, from which data are to be acquired and analyzed by a digital computer in an automatic fashion, has to present the data to the computer in binary digital form. Unfortunately, most instruments produce signals or data in an analog form. Analog signals are continuously varying and do not produce data in numbers. For example, the needle of a car speedometer moves in a continuous fashion from one end of the dial to the other as the speed of the car is increased from 0 to a certain maximum. The higher the speed of the car, the farther the needle moves. The digital (in numbers) information about the car's speed is derived from a scale printed on the dial. The scale printed on the dial is, in a way, manual digitization of an analog (continuous movement of the needle) signal. Now, if the speedometer has to be coupled to a digital computer to record the speed of the car automatically, some electronic device will have to be used which will digitize the information provided by the speedometer (movement of the needle). Such a device which automatically changes analog signals into digital (binary) signals is known as analog to digital convertor, or simply an ADC. Two important parameters of an ADC, accuracy and speed, are relevant for our purpose. Accuracy of ADC tells how close the numeric data is to the analog signal. Let us consider the example of the speedometer again and assume this time for simplicity that the minimum speed of the car is zero and the maximum speed 100 miles/hour. Now to read any speed in between these two extremes, the scale on the dial has to be divided into equal divisions. If there are only ten divisions, then we will be able to measure the speed of a car accurately only in steps of 10 miles/hour. When the scale is divided into 100 equal parts, the speed could be read in steps of 1 mile/hr. One thousand equal divisions will provide an accuracy of 0.1 mile/hr and so on.

The more divisions there are on the scale (in a given range), the better the accuracy. Similarly, in an ADC, a given range of signals is broken into divisions, the more divisions, the better its accuracy. The unit of divisions in the case of an ADC is a bit. A one bit ($2^1$) ADC divides a given range only into two equal parts, a two bit ($2^2 = 4$) ADC divides it into 4 equal parts, a three bit ADC ($2^3 = 8$) into 8 equal parts and so on. The more bits an ADC has, the better its accuracy. However, it takes more time to digitize a signal as the number of bits in an ADC increases. This brings us to the second parameter of an ADC, speed. The faster an ADC is, the higher rates of data it can digitize without any loss of information. Thus, speed and accuracy are inversely related. More accuracy means less speed and more speed means less accuracy.

In a scintillation camera, the X and Y signals produced by the position circuit are analog in nature and, therefore, have to be digitized before being recorded by a digital computer. The ADC's used for digitizing X and Y signals of a scintillation camera are 6 or 7 bit which means that the X and Y ranges are equally divided into $2^6 = 64$ or $2^7 = 128$ equal divisions respectively. In a scintillation camera the total range of X or Y signals will equal the diameter of the crystal. Therefore, each division of an ADC will correspond to either d/64 or d/128 cm of distance depending whether the ADC is 6 or 7 bit. For a 28-cm diameter (11″) crystal these values equal 0.44 and 0.22 cm respectively. Since the intrinsic resolution of a scintillation camera presently is in this range, greater accuracy is not needed. In terms of the speed of the ADC, it should be able to handle about 100,000 signals per second, because higher count rates are seldom encountered in nuclear medicine.

The digitization of X and Y signals into 64 or 128 divisions yields a $64 \times 64$ or $128 \times 128$ matrix. Thus, the analog image (which is two dimensional or areal distribution) is divided into $64 \times 64 = 4096$ or $128 \times 128 = 16384$ equal small areas known as pixels. A specific area on the crystal corresponds to a specific pixel and each pixel is assigned a specific location in the computer. Therefore, when a gamma ray interacts in the crystal, its pixel location is determined by the ADC's and a count is stored in the corresponding location in the computer. As more and more gamma rays interact, they are stored in the appropriate locations and finally a digitized image is formed. This is graphically illustrated in Fig. 11–8 using a $4 \times 4$ matrix instead of $64 \times 64$ or $128 \times 128$ which are used in practice.

The various areas where computer has helped tremendously in the practice of nuclear medicine are as follows:

***Automatic Acquisition of Images:*** In the case of static imaging, a digital computer is not as helpful as in the case of fast dynamic studies where images may have to be acquired every ½ second for a period of 100 seconds or more. In dynamic studies, a digital computer is the only practical way to acquire such a large amount of data accurately and efficiently. Another type of acquisition, known as multiple gate acquisition (MUGA) is also made practical by computer acquisition of data. Multiple gated acquisition is desired where the organ to be imaged moves in a periodic fashion, for example, heart. To acquire images in various phases of heart beat (such as systole or

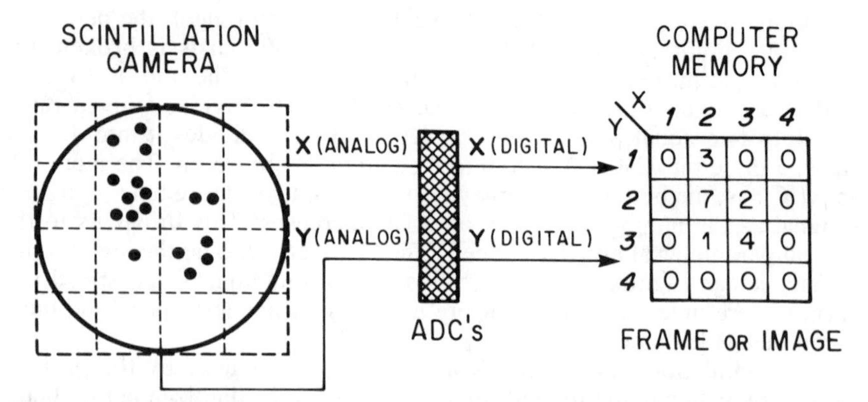


**Fig. 11–8.** Digitization of an image: Analog image produced by an Anger camera is divided into a number of (4000 or 16000) square cells or pixels. Counts from each pixel are stored in a separate location in a computer. Here, the X, Y ranges have been divided into 4 divisions. The resultant 4 × 4 matrix has 16 pixels.

diastole), the beat is divided into a number of time-segments, usually 16. The onset of the first time-segment is triggered by the R wave of an EKG monitor attached to the patient. Data recorded by the scintillation camera during each segment of time are stored by the computer in different parts of the memory. When the next beat starts, the data acquired in each segment of time of this beat are recorded (or added) in the corresponding locations of the data acquired during the first beat. This process is continued for a large number of beats, quite often up to 1000 or more. The reason for summing the data for each time-segment of a heart beat for so many heart beats has to do with the number of counts in the image of each time-segment (also called "gate"). Since a heart beat is generally of less than one second in duration, if one divides it into 16 time-segments, each time-segment is only 30 to 50 msec (depending upon the actual time of the beat). The number of counts detected by the scintillation camera during a 30 to 50 msec time interval, even for a 20 mCi $^{99m}$Tc radioactivity administration to the patient, is quite small. Therefore, by adding counts from a large number of beats for each time segment, images with sufficient numbers of counts (200,000 to 1,000,000) for each time-segment of heart cycle (phase) can be obtained.

***Display of Images:*** In scintillation camera imaging, even with a multi-imager, proper intensity setting of the oscilloscope light spot in every case is difficult. With computer acquired digital images there is no such problem, as images can be stored temporarily or permanently on a disc or magnetic tape and can be displayed over and over again with any desired intensity. The dynamic images can be displayed in a movie-like fashion so that the flow of a bolus of radioactivity can be followed. MUGA images of the heart can be displayed in a ciné mode which shows the motion of the heart during a heart beat.

***Analysis of the Images:*** The digitized images can be manipulated individually or added, subtracted, multiplied or divided by one another by the

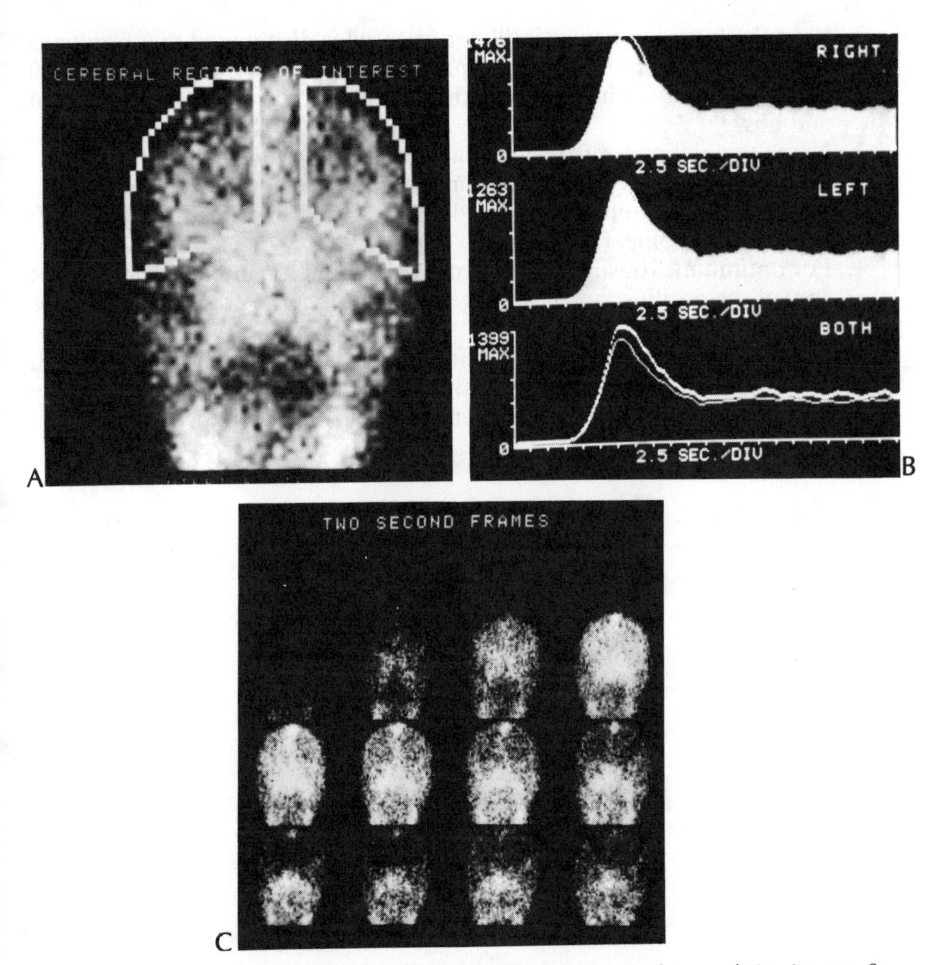


**Fig. 11–9.** Use of a computer to acquire, analyze and display nuclear medicine images. Sequential brain images of 1 second duration were acquired for 50 seconds. From these data, a composite image consisting of all 50, 1 sec images was formed, *A*. The areas of interest (left and right hemisphere) were chosen. Computer then formed dynamic curves showing time changes in the two regions of interest, *B*. Also shown, *C*, are first 16, 2 sec images which were formed by adding 2, 1 sec images successively.

computer to generate another image which may provide better appreciation of certain clinically important parameters. Also, counts can be determined and compared from areas of interest (ROI) in the same images or in images taken at different times. An example of the use of the digital computer in the case of a dynamic study is shown in Fig. 11–9.

## PROBLEMS

1. The crystal in a scintillation camera has an optimum thickness. What are the two properties that determine the optimum thickness?

2. If, in the future, a new scintillator material is discovered that produces four times the amount of light produced by NaI(Tl) crystal for a 140 keV gamma ray, what factor of the scintillation camera can be effected and how?
3. Name the collimator for which each of the following statements is true: (a) The field of view remains constant. (b) It magnifies the object. (c) Its field of view increases with distance and it minifies the object. (d) It inverts the object.
4. For optimum exposure of the film, how should the intensity of the light spot of the oscilloscope be changed if the number of counts in the image is increased?
5. An ADC is the most important link between a scintillation camera and a digital computer. If the resolution of scintillator camera improves by a factor of 10, what effect it will have on the desired accuracy and speed of the ADC?

# 12

# Operational Characteristics and Quality Control of a Scintillation Camera

A number of parameters of an imaging device play a major role in the delineation of a radioactive distribution. Of these, two, spatial resolution and sensitivity, are important. For scintillation cameras two other operational characteristics, uniformity and high count rate performance, have also to be considered.

These and some routine quality control procedures constitute the subject matter of this chapter.

## QUANTITATIVE PARAMETERS FOR MEASURING SPATIAL RESOLUTION*

Resolution is defined as the ability of a scanner to reproduce the details of radionuclidic distribution. The finer details an imaging device reproduces, the better resolution it has. How is resolution measured quantitatively? Two parameters—full width at half maximum (FWHM), and modulation transfer function (MTF)—have been used to measure resolution of an imaging device. Sometimes "bar phantoms" are used as a semiquantitative measure of resolution. These are discussed under quality control.

***Full Width at Half Maximum (FWHM):*** If we image a single point source and plot the intensity profile across its center, a bell-shaped curve similar to that shown in Fig. 12–1 results. This curve is known as the point-spread function of an imager. The full width of this curve at half maximum (FWHM) can be used to measure resolution quantitatively. A shorter FWHM implies

* Wherever the word "resolution" appears in this chapter, it should be read as "spatial resolution."

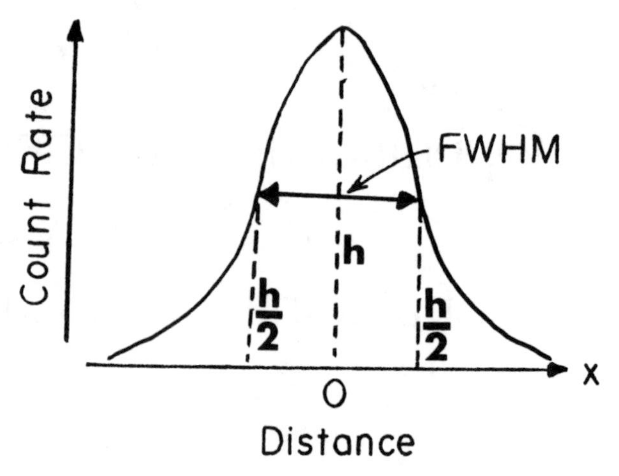


**Fig. 12–1.** Intensity profile through the center of a point source image with a scintillation camera. The width or narrowness of such a profile is a good measure of resolution. The narrower the curve, the better the resolution of the imager. The full width of this curve at the two points for which the response has decreased to one half of the maximum (FWHM) is customarily used as a quantitative parameter for measuring the resolution of an imaging device.

a better resolution of the imaging device. Measurement of FWHM requires a computer interfacing with a scintillation camera.

FWHM is a useful parameter for expressing the relationship of resolution to various parameters of an imaging device. Its main drawback is that it does not measure resolution under varying object contrast conditions. As a result it is possible to design two imaging devices which have equal resolutions according to this definition, although in practice one performs better than the other.

***Modulation Transfer Function (MTF):*** MTF gives the most complete characterization of the resolution of an imaging device, provided that the response of the imaging device is linear. Although the latter condition is not strictly met for imaging devices, the modulation transfer function is still useful in their evaluation.

To understand this parameter fully, knowledge of Fourier analysis is essential. However, to comprehend it conceptually, an analogy with sound is helpful. Any sound—the ding-dong of a bell or the pretty voice of a singer—is made up of a number of sound waves of different frequencies. Once the component frequencies and their strengths (amplitudes) are known for a given sound, it can then be resynthesized (in a laboratory) by proper superimposition of these frequencies. In a similar fashion, any spatial distribution (object) can be broken down into a number of spatial frequencies, and the original distribution (object) can then be resynthesized by the proper superimposition of these spatial frequencies.

How does this breaking up of a spatial distribution into component spatial frequencies aid in evaluating the imaging device? Not *per se*, but if we mea-

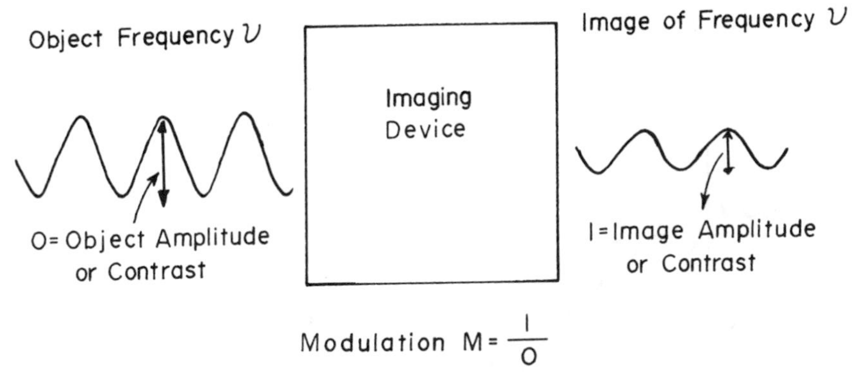


**Fig. 12–2.** Modulation transfer function (MTF). When a spatial frequency, $\nu$, of an object is imaged, its amplitude may change. The ratio of the amplitude in the image of a spatial frequency to that in the object is known as modulation M. Measurement of M as a function of $\nu$ yields the modulation transfer function (MTF) of an imaging device.

sure the degradation produced by an imaging device as a function of various spatial frequencies, the resulting function will provide the information needed to characterize the imaging system completely. The degradation (M) produced for a spatial frequency ($\nu$) by an imaging device is measured as the ratio of the contrast (amplitude of the wave) in the image frequency to the contrast in the object frequency (Fig. 12–2).

Measurement of M as a function of $\nu$, then, produces the MTF of the imaging device. When the value of M for a particular spatial frequency is 1, this indicates that there is no degradation of contrast for that frequency. If M equals zero, the imaging device is unable to reproduce this particular spatial frequency and, therefore, a value of zero represents the maximum degradation. Values of M between 1 and 0 represent the extent of degradation for a given frequency. An ideal scanner (which produces an exact image of an object) will have a value of M = 1 for all spatial frequencies.

Use of the MTF to compare the resolution of two different imaging devices can be appreciated easily by examining Fig. 12–3, which depicts the MTF for three imaging devices, A, B and C, respectively. Here, the MTF of the imaging device A is higher for all spatial frequencies than the MTF of the imaging devices B and C and, therefore, the imaging device A possesses the best resolution of the three imaging devices. The choice between B and C is difficult. At low spatial frequencies B is superior to C, whereas at higher frequencies C is superior to B. The selection of the particular imaging device in this case will depend on the type of objects to be imaged. If any object dominates in high frequencies, device C will be a better choice. If an object contains primarily low frequencies, then device B will be more advantageous to use. The MTF of an imaging device is difficult to measure directly. Instead, it is calculated from the line-spread function (discussed below) of a scanner which can be easily measured.

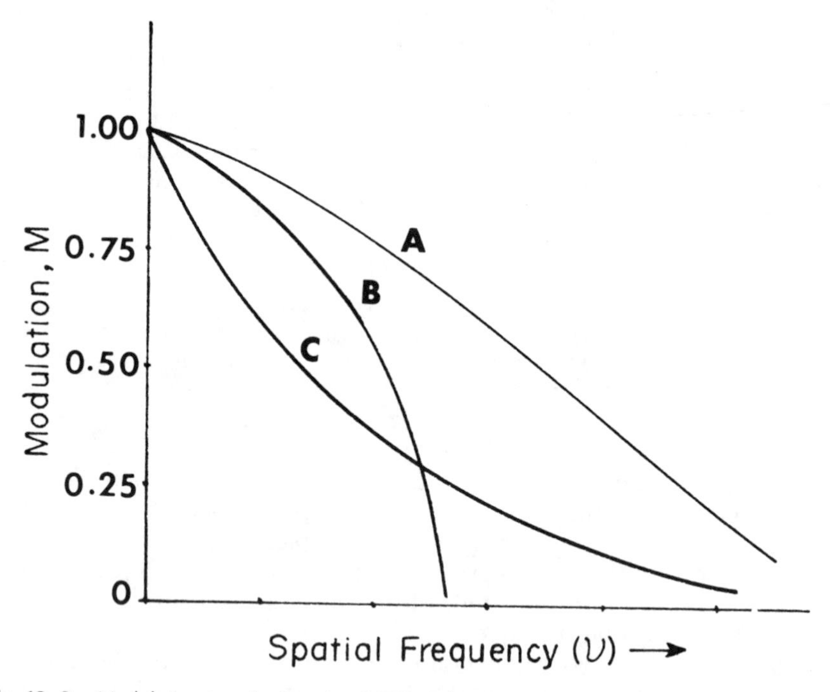


**Fig. 12–3.** Modulation transfer function (MTF) of three imaging devices. Modulation M for imaging device A is higher than that of B and C at all spatial frequencies. Therefore, it possesses the best resolution of the three. The choice between B and C is difficult because at low spatial frequencies B is superior to C, whereas at high spatial frequencies C is superior to B.

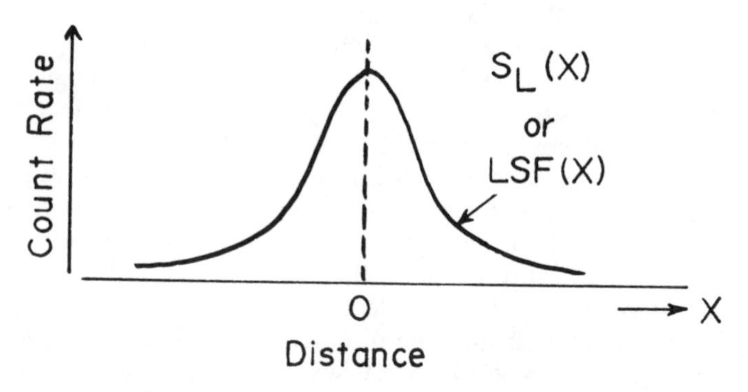


**Fig. 12–4.** Line spread function (LSF) of an imaging device. Using a gamma camera, LSF can be directly obtained from the image of a line source, provided it is interfaced with a computer system.

## QUANTITATIVE PARAMETERS FOR MEASURING SENSITIVITY

Besides resolution, the other important parameter of an imaging device is its sensitivity. Sensitivity can be defined as the ability of an imaging device to use efficiently all the photons which are available from an object within a given unit of time. Three parameters—point sensitivity, line sensitivity, and plane sensitivity—have been used to measure the sensitivity of an imaging device. Each has relative advantages and disadvantages.

***Point Sensitivity ($S_p$):*** This parameter is defined as the fraction of $\gamma$-rays detected per unit of time for a point source of radioactivity. In a scintillation camera, $S_p$ is more or less constant in the field of view of the collimator.

***Line Sensitivity ($S_L$):*** This parameter is defined as the fraction of $\gamma$-rays detected per unit of time per unit of length of a very long line source of uniform radioactivity. The count profile of a line source, as determined by an imaging device, through a direction perpendicular to the line source is known as the line-spread function (LSF) (Fig. 12–4). LSF(x) is primarily used in the calculation of the MTF of an imager as follows:

$$\text{MTF}(\nu) = \frac{\int_{-\infty}^{\infty} \text{LSF}(x)\cdot\cos(2\pi\nu x)\cdot dx}{\int_{-\infty}^{\infty} \text{LSF}(x)\cdot dx}$$

***Plane Sensitivity ($S_A$):*** Plane sensitivity is defined as the fraction of $\gamma$-rays detected per unit of time per unit of area of a large-plane source of uniform radioactivity. This parameter is commonly used to compare the sensitivities of two imaging devices. The principal advantage of $S_A$ is the ease with which it can be measured. Plane sensitivity does not vary with the distance of the plane source from the collimator, so long as the area of the plane source is larger than the field of view of the collimator at that distance.

## FACTORS AFFECTING SPATIAL RESOLUTION AND SENSITIVITY OF AN IMAGER

The resolution and sensitivity of an imaging device depend on a number of variables described below. Theoretically, the exact relationships of these variables to resolution and sensitivity are difficult to obtain in a generalized case. However, by making certain assumptions, these relationships can be expressed in simplified mathematical form. Using these formulas, an approximation of the dependence of resolution and sensitivity on a given variable can be easily deduced. In the following discussion we shall be using these simpler formulas. In addition, we shall employ R and $S_A$ as a measure of the resolution and sensitivity, respectively, of an imaging device, assuming that (1) there is no septal penetration by gamma rays in the collimator of the scanner, and (2) there is no scattering of $\gamma$-rays in the radionuclide source.

### Scintillation Camera

The resolution R of a scintillation camera is comprised of $R_1$ and $R_2$, where $R_1$ is the intrinsic resolution of a scintillation camera and $R_2$ is the resolution of the collimator used with the scintillation camera. Resolution, R, is approximately related to $R_1$ and $R_2$ as follows:

$$R = \sqrt{R_1^2 + R_2^2} \qquad [3]$$

The intrinsic resolution $R_1$, which is a measure of the uncertainty in the localization of the point where light is produced in the crystal, is degraded with an increase in the thickness of the NaI(Tl) crystal, and is improved with an increase in $\gamma$-ray energy. The improvement in the intrinsic resolution of a camera with the gamma ray energy is shown in Figure 12–5.

Resolution $R_2$ depends on various collimator parameters such as collimator length, L, diameter of holes, d. We shall limit our discussion here to a parallel-hole collimator (Fig. 12–6), although similar considerations apply for a converging or diverging collimator. $R_2$ in this case depends on the hole diameter d, length of the collimator L, thickness of the crystal c, and the distance F of the source from the collimator face. $R_2$ is given by the following expression:

$$R_2 \simeq \frac{d(F + L + c)}{L} \qquad [4]$$

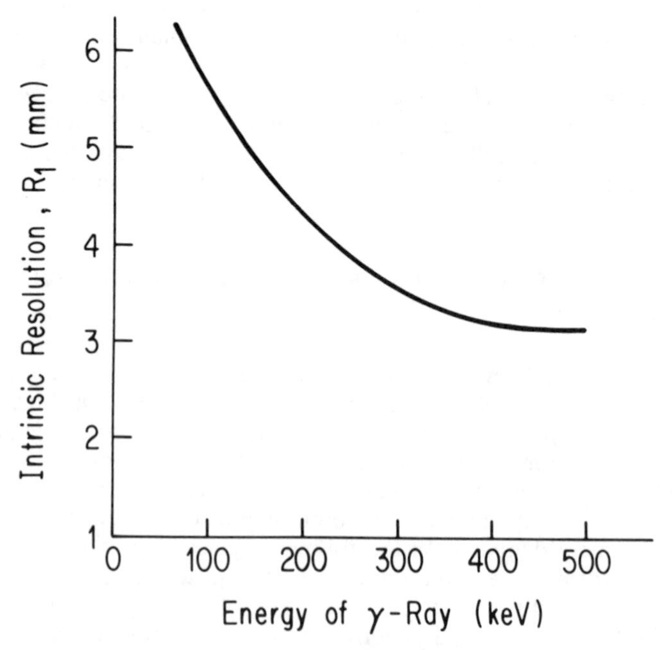


**Fig. 12–5.** Dependence of intrinsic resolution ($R_1$) of an Anger camera on $\gamma$-ray energy.

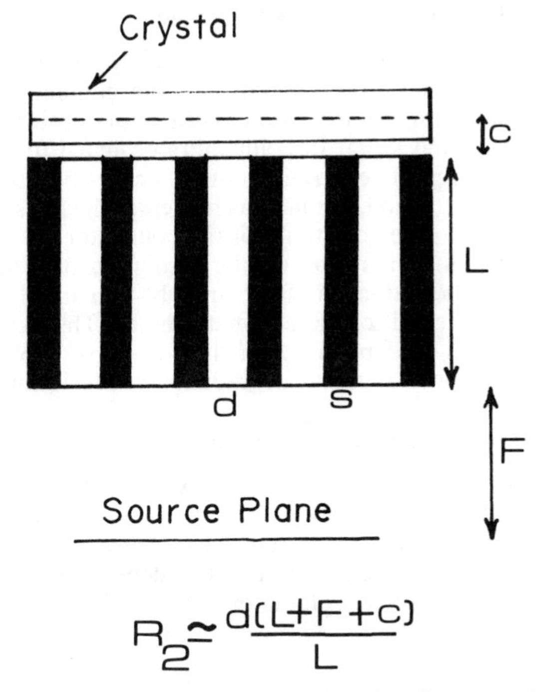


**Fig. 12–6.** Dependence of resolution, $R_2$, of a parallel-hole collimator on the length, L, of the collimator and the diameter, d, of the holes. Reducing d or increasing L improves $R_2$. Sensitivity of such a collimator also varies as inverse square of the resolution.

When the source is at a distance F from the collimator, to improve $R_2$ we have to either reduce d and c, or increase L. The sensitivity $S_A$ depends on d, L, D, $\epsilon_p$ and s, as follows:

$$S_A \simeq \frac{\pi d^4}{64L^2} \cdot \frac{3D^2}{4(d + s)^2} \cdot \epsilon_p \qquad [5]$$

It can be seen from equation 5 that $S_A$ can be increased by either increasing d or decreasing L, the opposite of that required to improve the resolution $R_2$ (for an optimum collimator, $S_A \alpha R_2^2$). Therefore, a twofold gain in $R_2$, either by increasing d or decreasing L, necessitates a fourfold sacrifice in $S_A$. Sensitivity of a parallel-hole collimator can be increased by increasing the diameter D of the crystal ($S_A \alpha D^2$). Accordingly, a 13″-diameter crystal scintillation camera has a sensitivity equal to $\frac{13^2}{11^2}$, or 1.4 times that of an 11″-diameter crystal scintillation camera. This gain in sensitivity of a large crystal scintillation camera is of value only in cases where the organs to be imaged are of the same size as the crystal.

### Loss of Spatial Resolution Resulting from Septal Penetration

In the previous discussion we assumed that no $\gamma$-ray could reach the radiation detector by penetrating the septa of the collimator. Such an assumption is justifiable only in the case of collimators designed for use with low-energy $\gamma$-rays (<150 keV). In collimators designed for use with high-energy $\gamma$-rays, some septal penetration always occurs because a further reduction of septal penetration, by increasing septal thickness, will produce an unacceptable loss in the sensitivity of the collimator.

The effect of septal penetration on the resolution of a collimator is to degrade it. Septal penetration, in effect, may be construed as an effective increase in the diameter, d, of the collimator holes. The extent of the degradation of resolution is dependent on the degree of septal penetration. Higher penetration leads to greater degradation of resolution.

### Variation in Spatial Resolution with Depth

Resolution of a scintillation camera varies with the depth or distance from the face of the collimator. For the scintillation camera the best resolution is achieved at the face of the collimator. The farther away from the face of the collimator, the poorer the resolution. The dependence of resolution on depth of a typical parallel-hole collimator of a scintillation camera is shown in Fig. 12–7 where both R and MTF were used as the index of resolution.

## UNIFORMITY AND HIGH COUNT RATE PERFORMANCE OF A SCINTILLATION CAMERA

Besides resolution and sensitivity, two other characteristics of a scintillation camera, uniformity and high count rate performance, are also important for the optimal operation of a scintillation camera.

***Uniformity:*** It is the ability of a scintillation camera to reproduce a uniform radioactive distribution (mind you, in a uniform source, there are no details or variations of count rate and, therefore, resolution has little to do with the uniformity response of a scintillation camera). In practice, all scintillation cameras produce non-uniform or inhomogeneous images of a uniform source to varying extent. These inhomogeneities or the count rate variations from one area to another in the image of a uniform source may be in amounts up to ± 10%. In the image itself, these areas of increased or decreased activity show as "hot" or "cold" spots respectively (Fig. 12–8).

Even though some inhomogeneity in the image of a uniform source is caused by slight variations in the thickness of the NaI(Tl) crystal and in the transmission of the gamma rays by the collimator, the dominant cause of non-uniform response is electronic in nature. It is related to differences in response of PM tubes and to the difference in transmission of light produced at different places in the crystal. These differences cause the mispositioning of some of the counts. A good demonstration of mispositioning of some counts occurs in scintillation cameras occasionally when a straight line radioactive source is imaged. The image of a linear radioactive source appears

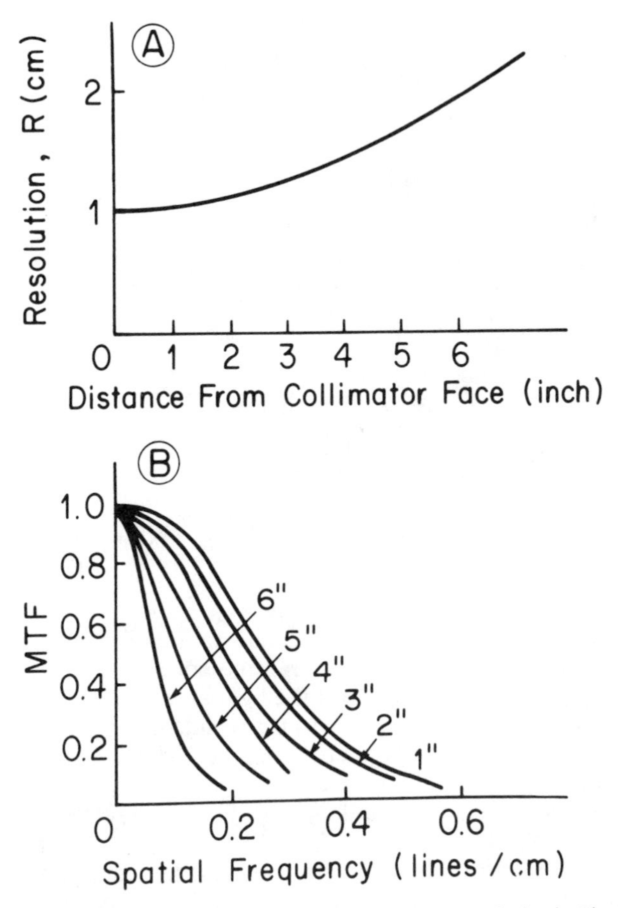


**Fig. 12–7.** Variation of the resolution of a scintillation camera with depth. The best resolution is obtained closest to the collimator. In *A*, R has been used as a measure of resolution; in *B*, MTF has been used for this purpose. Both, however, show similar behavior in variation of resolution with depth. The number on the MTF curves in *B* gives the distance from the collimator face.

as an arc which may be curved in or out from the center of the scintillation camera (pin cushion or barrel distortion in optical analogy). Even though, such distortions of linear sources are visible only when the scintillation camera is not properly "tuned." A certain amount of non-linearity always persists even when it is not discernible on visual inspection of the image and the camera is properly tuned. These small non-linearities result in a visible non-homogeneous response of the scintillation camera to a uniform source. To keep the non-homogeneities to a minimum, scintillation cameras must be properly tuned. Tuning involves the readjustment of individual PM tube gains so as to overcome the individual differences in various PM tube responses. Since the PM tube gain may drift because of fluctuations in voltage etc., it is essential that uniformity of response be checked routinely.

Another aspect of the non-uniform response of a scintillation camera is

**Fig. 12–8.** Uncorrected "flood" showing the non-uniform response of a scintillation camera. The bright outer ring in the image is due to edge packing. In new cameras, this ring is no longer visible.

known as edge packing. This is also a manifestation of mispositioning of the counts. It appears as a bright ring around the edge of an image (Fig. 12–8) and is caused by internal reflection of the light at the edge of the crystal and the fact that the PM tubes are present on one side only. As a result, counts coming from near the edges are bunched together. The region of edge packing is never used in clinical studies and, therefore, it is always masked by a lead ring around the collimator. Thus, the useful field of view (UFOV) is always smaller than the crystal size.

In modern scintillation cameras (so-called digital cameras), the non-uniformity of response is improved considerably by addressing its causes—local variations in the amount of light transmitted to a PM tube and the non-linear response of the X,Y circuit due to slightly different PM gains. These non-linearities are very carefully measured and, from these measurements, a correction matrix formed. This correction matrix then appropriately re-positions on line all the subsequent counts detected by the camera using a microprocessor. As a result of these corrections, the uniformity of the scintillation camera has tremendously improved and is 2% or better in the useful field of view (Fig. 12–9). Moreover, the electronic components and the PM tubes used in these cameras are more stable than those employed in older cameras. Therefore these generally do not require tuning as often; a monthly or even a quarterly interval is adequate for routine studies. However, the measurements of the correction matrix are too complicated to be done by in-house personnel and, therefore, if a camera detunes significantly from its prescribed limits, a service call is necessary.

The non-uniform transmission of gamma rays by the collimator is tested separately. If it is found unacceptable, the collimator is replaced. Once a satisfactory collimator is found, occasional physical inspection for the presence of dents or unusual marks assures the integrity of the collimator.

**Fig. 12–9.** Microprocessor—corrected "flood" showing more uniform response of a scintillation camera.

***High Count Rate Performance:*** Since a gamma camera employs a NaI(Tl) crystal to detect gamma rays as well as to determine the location of gamma ray interaction in the crystal, at high count rates, besides loss of counts due to the finite dead time, mispositioning of counts also takes place. At high count rates, the probability of two gamma rays simultaneously interacting in the crystal (within the dead time of the detector) increases sharply. If one or both gamma rays interact through the photoelectric effect, the total light produced will be more than the light produced if only one gamma ray interacted that way. Therefore, if the PHS is set on the photopeak, both of these gamma rays, mimicking as one but with higher energy, will be rejected by the PHS. On the other hand, if both gamma rays interact through the compton effect and each interaction produces enough light so that the sum of the two is equal to the light produced by the photoelectric interaction of one gamma ray, the two gamma rays, mimicking as one but with correct energy, will be accepted by PHS. However, the location of the interaction will be the mean of the locations of the interactions of the two gamma rays in the displayed image. Thus, a misposition of the count occurs.

The dead time of a scintillation camera is made up of both components, paralyzable and non-paralyzable. Under ideal conditions it is between 1 to 2 $\mu$sec. However, the dead time of a scintillation camera is a complex function of window width used, scattering material around the source, presence of more than one gamma ray in the emissions of the radionuclide used, etc. Therefore, it is essential to determine the dead time of a scintillation camera under conditions typically employed in clinical situations. In these circumstances, dead time may be as much as 10 to 15 $\mu$sec. Count rate performance of a typical scintillation camera is shown in Figure 12–10.

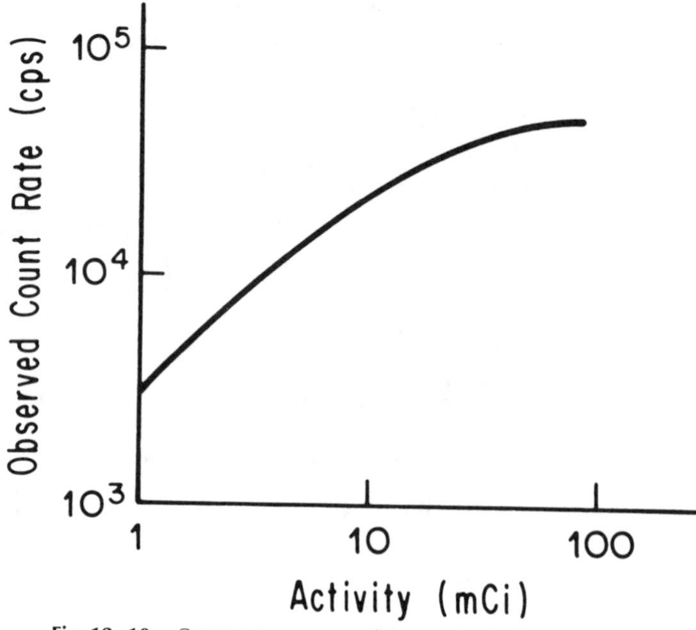


**Fig. 12–10.** Count rate response of a typical scintillation camera.

## QUALITY CONTROL OF IMAGING DEVICES

Correct interpretation of an image depends upon the accuracy of the data in the image. Therefore, it is of utmost importance that nuclear imaging devices operate optimally and reliably. Since these instruments employ a variety of electronic components whose response may be affected by changes in line voltage or ambient conditions, the best way to insure accuracy and reliability of these instruments is through a continuing program of quality control. The quality control procedures are intended to monitor day to day variations in some of the performing characteristics so as to alert us, in time, of the malfunction or unacceptable behavior of these instruments. These are semiquantitative procedures and, therefore, do not measure the operating characteristics of an imaging device accurately or completely. The procedures given in the following sections are brief and therefore illustrate only the important aspects of such an endeavor.

***Scintillation Camera:*** The three most common parameters routinely tested to assure maximal performance of a scintillation camera are (a) peaking, (b) field uniformity and (c) spatial resolution.

*Peaking:* This test, which is performed every day, assures that the window of PHS is correctly set on the desired photopeak. The following steps are involved in peaking a scintillation camera.

1. Place a small radioactive source under the scintillation camera (if the

collimator is removed, 100 to 200 $\mu$Ci of $^{99m}$Tc radioactivity is enough; with the collimator on, 1 to 2 mCi of $^{99m}$Tc radioactivity is preferred).

2. Place the camera in Spectrum mode, and set the energy and window for the radionuclide being used (for $^{99m}$Tc, 140 keV and 20% window).

3. Observe whether the photopeak is within the window or not. If not, change the high voltage slowly so as to center the photopeak in the window.

4. Take a picture of the spectrum and record the high voltage setting. The high voltage setting should not change more than 10% from one day to the next. If the change is greater, investigate the cause because this may lead to a malfunction. At this point also note that if there are any fingerprints, dirt, etc. on the oscilloscope screen.

*Field Uniformity:* This test, which should be performed daily, assures that the count rate variations in the image of a uniform source are within an acceptable range and the scintillation camera is properly tuned. Since the collimator response is not expected to change unless there is evidence of physical damage, only the intrinsic field uniformity is checked.

The following steps are involved:

1. Remove the collimator, and move the detector head 4 to 5 feet above the floor.

2. Place a 100 to 200 $\mu$Ci small volume (0.2 ml) source of $^{99m}$Tc on the floor. (Have absorbent paper under the source to avoid contamination of the floor.)

3. Set the preset count of $10^6$ counts for a small field of view and $2 \times 10^6$ counts for a large field on the view scintillation camera.

4. Adjust the display oscilloscope intensity to the $10^6$ or $2 \times 10^6$ count level depending upon the type of scintillation camera.

5. Turn the scintillation camera on and record the image.

The resultant image is visually evaluated for uniformity, image shape (it should be a nice circle) and other artifacts. Variations greater than $\pm 10\%$ are easily detectable in the image and are unacceptable. On the type of scintillation camera with uniformity correction, two images are recorded, one with the microprocessor off and the other with the microprocessor on.

*Spatial Resolution (Intrinsic):* Since measurement of the two quantitative indices of spatial resolution, MTF and FWHM, is time consuming, a semi-quantitative evaluation using a 90° quadrant bar phantom is performed. The phantom consists of four sets of parallel lead bars arranged in four quadrants of a lucite holder as shown in Figure 12–11. The spacings and width of the lead bars vary between each quadrant but are the same within each quadrant. The smallest spacing is chosen to be smaller than the spatial resolution of the scintillation camera. The procedure for evaluation of the intrinsic spatial resolution is identical to that of field uniformity (steps 1 to 5) except, under step 1, after removing the collimator, the bar phantom is attached in front of the crystal of the scintillation camera. The resultant image is inspected for the separation of the finest bar spacing as well as linearity of the bars. The recommended frequency for this test is weekly.

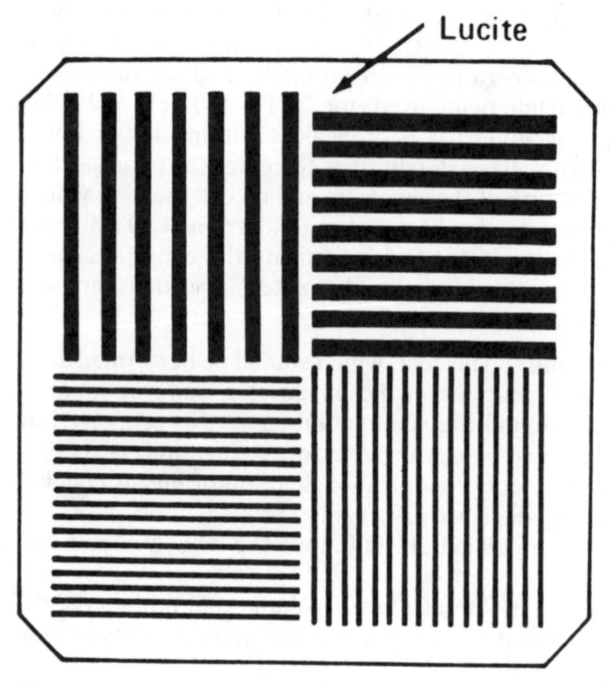


**Fig. 12-11.** A 90° quadrant, parallel lead bar, phantom used for quality control of the spatial resolution of a scintillation camera.

## PROBLEMS

1. The resolution and sensitivity of a scintillation camera are 10 mm and 50,000 counts per minute per microcurie respectively. What will be the sensitivity of this camera if a new collimator with 5 mm resolution is used instead?
2. Why are there so many indices of resolution? Give the best application for each of them.
3. Determine the overall resolution of a scintillation camera if the resolution of the collimator is 10 mm and the intrinsic resolution is (a) 10 mm, (b) 5 mm, and (c) 1 mm.
4. Why is about 1 inch of crystal at the edge not usable for imaging in a scintillation camera?
5. Besides the expected loss of counts, what problem does the dead time of a scintillation camera cause that does not occur in other counting situations?
6. How often (every day, weekly, or yearly) should the following parameters of a scintillation camera be checked? (a) energy resolution, (b) uniformity, (c) spatial resolution, (d) sensitivity and, (e) energy calibration.

# 13

# Detectability or Final Contrast in an Image

The primary goal in imaging is to detect the smallest possible localized or focal abnormalities (lesions) that may be present in an organ. Because of practical limitations, however, it is not possible to attain this goal. There is a lower limit of detectability below which a lesion cannot be visualized. This limit is determined by a number of parameters which are discussed below.

## OBJECT CONTRAST

The main purpose of any imaging device is to record the details of an object faithfully in its image. What do we mean by the details of an object? These are the spatial variations of a given parameter such as light intensity in photography, transmitted x-ray intensity in diagnostic radiology, and the concentration of radioactivity in scanning. Such a parameter, in the language of the physicist, is known as object contrast, and is most important in the detectability of a lesion. In nuclear medicine the object contrast is created in the organ of interest by the use of radiopharmaceuticals which either selectively localize in the abnormal tissue as compared to the normal tissue, or vice versa. In either case, the higher the variation between the concentration of radioactivity in the normal and abnormal tissue, the easier it is to detect an abnormality. Therefore, radiopharmaceuticals which produce greater contrast in the lesion have better detectability compared to those which produce smaller contrast.

For quantitative purposes we may define the object contrast, $C_0$, as follows:

$$C_0 = \frac{\text{concentration in abnormal tissue} - \text{concentration in normal tissue}}{\text{concentration in normal tissue}}$$

When there is no differential (variation) between the concentration of radioactivity in the abnormal and normal tissue, $C_0 = 0$ (*i.e.*, there is no contrast). Such a radiopharmaceutical producing zero object contrast will be of no use in the detection of abnormal lesions. When the concentration of the radioactivity is higher in the abnormal area than that of the normal area, then $C_0 > 0$, and the radiopharmaceutical produces a positive contrast. For example, radiopharmaceuticals currently used for brain scanning produce positive contrast, with values of $C_0$ generally ranging from 15 to 25. In cases where there is less radioactivity in a lesion than in normal tissue, radiopharmaceuticals produce a negative contrast ($C_0 < 0$). The value of $C_0$ for negative contrast cannot be increased more than $-1$, a value indicating that there is almost no radioactivity in the lesion. For example, radiocolloids used for liver scanning are preferentially localized in normal tissue with almost no radioactivity in the abnormal lesions. Since radiopharmaceuticals producing positive contrast can achieve higher contrast values ($C_0 \geq 1$) than those producing negative contrast, the former potentially have far better detectability than the latter.

## SPATIAL RESOLUTION AND SENSITIVITY OF AN IMAGING DEVICE

An imaging device, such as a scintillation camera, registers the details of the distribution of a radionuclide as a photographic camera records the details of an object or scene. In both cases, a physical device is used to form an image of an object. This is true, in fact, of any imaging process using such diverse instruments as an electron microscope, telescope, or x-ray tube. In all cases, the aim is to reproduce exactly the object contrast in the image.

Unfortunately, no imaging device is capable of reproducing all the details of an object in an image and a certain loss of detail (object contrast) is inevitable. The parameter of an imaging device which characterizes the extent of the loss of object contrast, or measures the faithfulness or the fidelity for reproduction of object contrast, is called spatial resolution. An imaging device which possesses better spatial resolution is capable of reproducing finer details of an object (smaller object contrasts) and is able to detect smaller lesions than an imaging device with a poorer spatial resolution. Therefore, the spatial resolution of a scanner is an important parameter which strongly influences the detectability of a lesion.

What limits the spatial resolution of an imaging device? Theoretically, there are no limitations in designing a scanner with fine spatial resolution capabilities. The limitations arise from the two practical constraints: (1) the radiation dose to the patient has to be kept low and (2) the time of scanning should be reasonably short.

The effect of these two restrictions is to limit the number of $\gamma$-rays that can be detected and displayed in an image. As we shall see in the following section, the total number of $\gamma$-rays (photons) in an image is also an important parameter affecting the detectability of a lesion. To obtain a given number of photons in an image within a limited time, therefore, the imaging device should possess high sensitivity. The sensitivity of an imaging device is a measure of its ability to detect $\gamma$-rays efficiently. A more sensitive device will require a shorter interval of time to detect the same number of $\gamma$-rays than a less sensitive device.

Unfortunately, the sensitivity of an imaging device is related as the inverse square to its spatial resolution (see Chapter 12). Therefore, an imaging device with a spatial resolution better by a factor of 2 than a given imaging device will have a fourfold loss of sensitivity. This loss of sensitivity theoretically necessitates either a fourfold increase in the scanning or imaging time or a fourfold increase in the radiation dose to the patient. In actuality the cost to improve the spatial resolution by a factor of two will be more than fourfold, as will be seen in the following section.

## STATISTICAL NOISE OR INFORMATION DENSITY

For reproducing the details of an object, even with high contrast, an imaging device with good spatial resolution is not, in itself, enough. The human eye is capable of seeing minute details of a well-lit object, but fails to perceive even large objects in a dark room. Hence, the amount of available light (number of photons) is another important parameter affecting the visualization of the details or contrast of an object.

This point is strikingly demonstrated in Fig. 13–1, showing six photographs of a pretty girl taken in succession with increasing numbers of total photons. With a smaller number of photons, only large details with high contrast are visible. As the number of photons increases, finer details of the object become visible in the image. In other words, the number of photons needed to visualize a given detail in an object is related to its contrast in the object. This relationship between the number of photons and the object contrast is not limited only to photography but is a generalized phenomenon of any imaging process.

Consequently, the net contrast in an image (it is the contrast in the image which allows us to determine the presence of a lesion) is derived from two components: the finite spatial resolution of the imaging device and the number of photons which make up the image. The latter component determines statistical noise in the image. In a well-lit object (a large number of photons) the limiting factor for reproducing the object contrast is essentially the spatial resolution of the imaging device, whereas in a dark room (hardly any photons) the limiting factor is the statistical noise in the image. In a moderately-lit room (a limited number of photons) the final contrast in the image is determined by both.

In nuclear medicine the statistical noise of an image is related to information density (ID), which is defined as the number of $\gamma$-rays detected per $cm^2$ of an object. If we multiply the information density by the area of the object, we obtain the average number of photons in an image. When performing a scan with a rectilinear scanner, it is convenient to set the information density; in the scintillation camera, it is easier to set the total number of photons. In both cases the statistical noise can be reduced by increasing either the information density or the total number of photons in the image. The statistical noise or error of N counts in an image equals to $\sqrt{N}$ (p. 27).

Since the total number of photons or information density needed to vis-

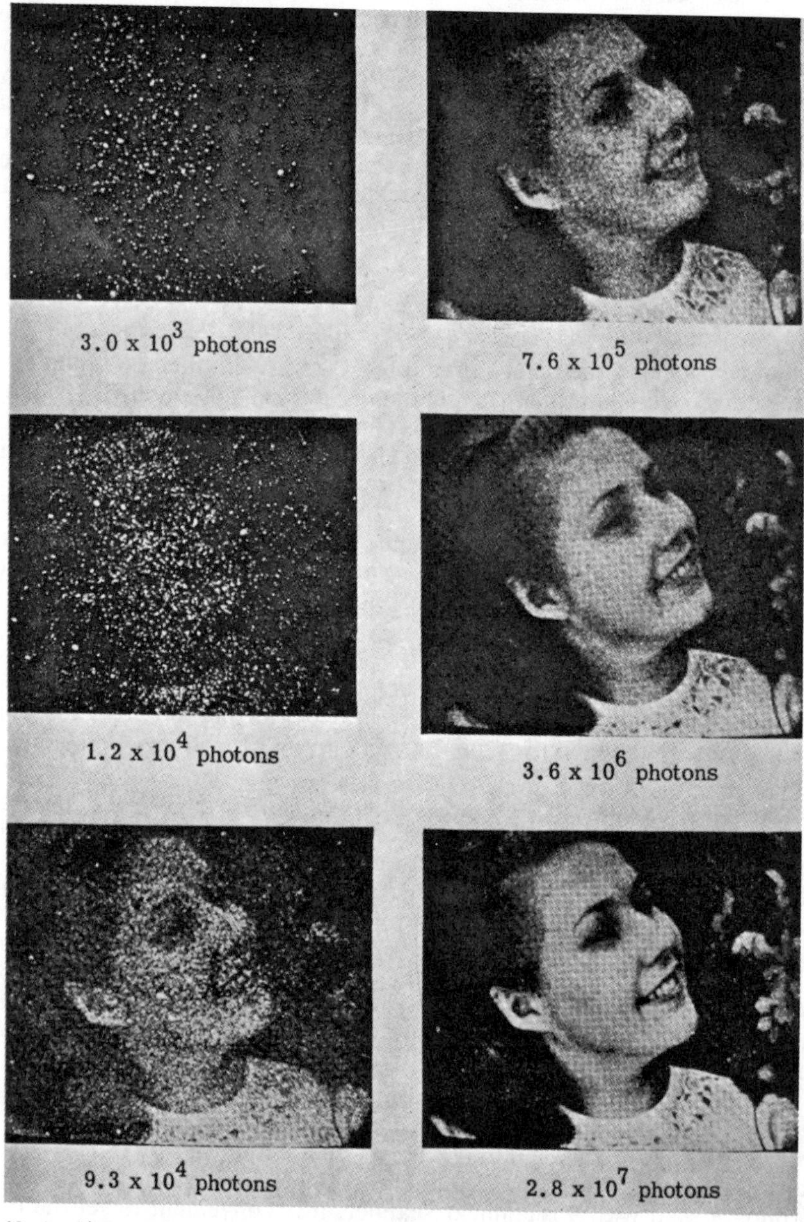


**Fig. 13–1.** Photographs taken under identical conditions except for variations in the total amount of light (number of photons). As the number of photons increases, greater detail is visible. A photographic camera possesses fine spatial resolution; yet, without an adequate number of photons that resolution is useless in providing the finer details in an object. (From Rose, A.: *J. Optical Society of America, 43*:715, 1953.)

ualize a lesion depends upon its contrast in the object, an information density of approximately 1,000 counts/cm$^2$ of the organ (or approximately 300,000 counts in routine brain scans) is generally accepted as an optimum value for currently available scanners. Use of information densities much lower than this value may cause border-line lesions (with low contrast) to be overlooked. On the other hand, increasing the information density higher than this value will not significantly affect the detectability because, at this level, the limiting factor for detectability is the resolution of the imaging device.

In the future, if imaging devices with improved spatial resolution but without any loss of sensitivity become available, higher information density will have to be employed. For example, an improvement in spatial resolution by a factor of 2 will necessitate an increase in information density approximately by a factor of 4. For the present day imaging devices, however, when we couple this fact with the fact that improving the spatial resolution will reduce the sensitivity by a factor of 4, we are talking about a factor of roughly 16: that is, an improvement in the lower limit of detectability by a factor of 2 using present-day imaging devices and radiopharmaceuticals will lengthen the time of the study approximately 16 times.

## PROJECTION OF VOLUME DISTRIBUTION INTO AREAL DISTRIBUTION

An imaging device, such as a gamma camera, in essence integrates the radioactivity present in the third dimension (depth) of an organ. The result of such an integration (*i.e.*, the weighted sum of radioactivity with depth) is to degrade the detectability of a lesion by lowering the object contrast (see Fig. 13–2 for a hypothetical situation).

## COMPTON SCATTERING OF $\gamma$-RAYS

As explained in Chapter 10, a $\gamma$-ray originating outside the field of view of a collimator can still reach the radiation detector as a result of compton scattering. The effect of such scattering is to reduce the object contrast and, therefore, to degrade the detectability of a lesion. Pulse-height selection used to discriminate against scattered $\gamma$-rays has proven quite effective in restoring some of the object contrast lost as a result of compton scattering. However, as the energy resolution of NaI(Tl) detector is not very high, a certain amount of object contrast is still lost. Also, as one narrows the window to reject the unwanted (scattered) counts, some of the good counts are also rejected. Therefore, too narrow a window may improve the image contrast somewhat, but it will be at the cost of sensitivity. This will not be the case for detectors with better energy resolution such as Ge(Li) detector. However it has other shortcomings as was pointed out in Chapter 8.

## ATTENUATION

Since attenuation of $\gamma$-rays by the tissue of distribution as well as by the overlying tissue reduces the number of $\gamma$-rays reaching a scanner, such a

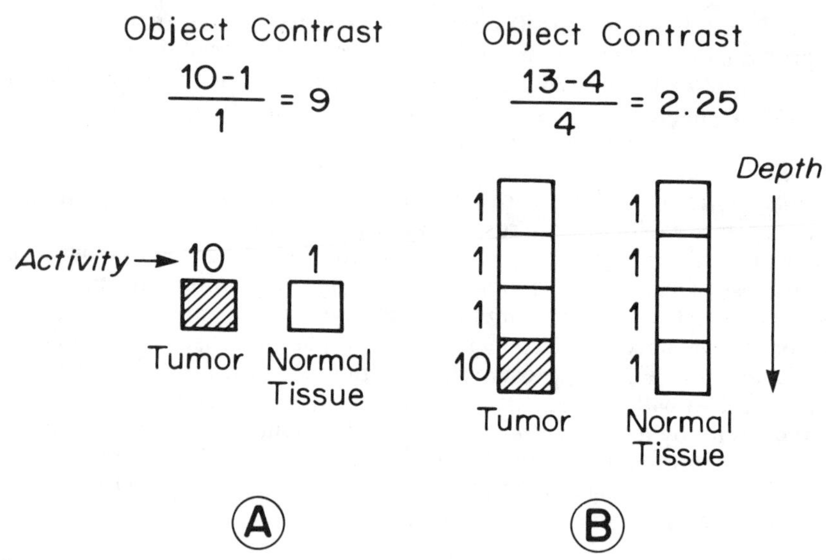


**Fig. 13–2.** Loss of object contrast as a result of the weighted sum in relation to depth of the radioactivity in routine area scanning. In this example, as a result of summation, a tumor with a contrast of 9 appears to have a contrast of 2.25 only. Here equal weight is given to each box; in actual scanning their exact contribution to summation will depend on the variation of the sensitivity with depth.

reduction adversely affects the detectability of a lesion. As a result, lesions that are on the surface of an organ are easier to detect than those lying deep within the organ.

## MOTION OF THE OBJECT

Motion of the patient or the organ (particularly the liver, lung and heart, which can move as much as 2 cm) also affects the detectability of a lesion in scanning. The effect of motion in an object is to reduce the object contrast which then results in a diminished contrast in the image. By using a computer with a scintillation camera it is feasible to eliminate some of the loss of object contrast due to motion. Since the clinical usefulness of such a procedure, compared to the time and effort involved, has not been convincingly established, we shall not go into the details of this method.

## DISPLAY PARAMETERS

The ability to visualize an abnormality is also affected by the display parameters such as exposure settings, the type of film employed (H-D curve of the film) etc. For the proper display, the count rate variations of interest should match the H-D curve's useful latitude (between points A and B of Fig. 10–7).

## OBSERVER VARIATIONS

In noise limited images, the detectability of a lesion is also dependent on the observer (in this case, the nuclear medicine physician). Since in noise limited images, a decision has to be made whether a given lesion is a manifestation of noise or not (false or true) such a decision may vary from observer to observer. Observation variation can be studied using what is known as receiver operated characteristics (ROC) curves.

## PROBLEMS

1. What should be the minimum contrast between a normal liver and a metastatic lesion (2 cm in diameter) for it to be visualized (at 99% confidence level) when the information density of a liver scintigram is 1000 counts/cm$^2$?
2. Do the radiopharmaceuticals designed to study blood flow in the brain have more or less contrast than the radiopharmaceuticals designed for tumor detection? Which of the two has better detectability?
3. Disregarding the consideration of cost, suggest a way to increase the sensitivity and therefore the detectability of a scintillation camera.
4. Why is the detectability for tumor detection in brain similar in nuclear medicine and computed tomography when the resolution for computed tomography is an order of magnitude higher than that in nuclear medicine?

# 14

# Emission Computed Tomography

Area scanning, even with multiple views, does not provide accurate three dimensional information concerning radionuclide distribution. Also, since a scanner more or less integrates the information from the third dimension (depth), area scanning of a three dimensional distribution results in lower contrast of lesions (Chapter 13). With the advent and success of computed tomography (CT) in diagnostic radiology, similar concepts and techniques have been applied in nuclear medicine as Emission Computed Tomography (ECT). Tomography, in general, is divided into two categories, transverse and longitudinal. It is the transverse tomography which has found success in routine clinical and research use. Transverse tomography can be performed with radionuclides emitting single photons (x- or $\gamma$-rays) or with radionuclides emitting positrons. Transverse tomography with single photons is commonly known as SPECT, whereas with positron emitters it is known as PET. Longitudinal tomography is performed with single photons only.

This chapter describes the principle of transverse tomography, instruments commonly used for SPECT and PET, and the relative merits of the two types. The mathematical techniques used in reconstruction of a tomograph are highly sophisticated and are beyond the scope of this book. Therefore, I shall limit myself to a phenomenological description only. Longitudinal tomography is also discussed briefly.

## PRINCIPLE OF TRANSVERSE TOMOGRAPHY

In its simplest form, a detector acquires data from a thin axial section containing radioactivity by linear scanning from multiple directions around the cross section as shown in Fig. 14–1. In its complex form and in order to reduce data acquisition time, data may be acquired from multiple thin cross sections and multiple directions simultaneously by using a large number of detectors or a scintillation camera. In either case, the principle of transverse tomography is the same and consists of two steps: acquisition of linear projection data of a thin cross-section from multiple directions and reconstruction of the cross section from this data.

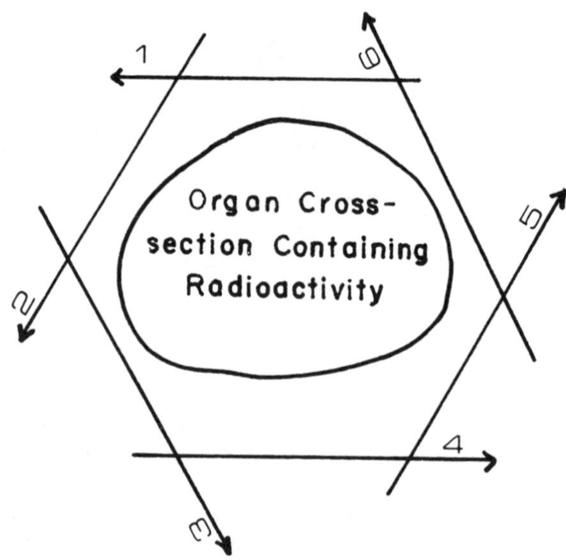


**Fig. 14–1.** Principle of axial tomography. A thin cross-section of an organ is scanned from several tangential directions (in this example, six). From these scans, the cross-sectional radioactive distribution is reconstructed, using a variety of superimpositional techniques.

***Considerations in Data Acquisition:*** Let us consider a simple case where a single detector linearly scans a cross section from many directions. Let us also assume that as the detector moves from one position to the next along a line, it receives counts from columns which are perpendicular to the direction of a scan and chosen to be equally spaced. This is shown in Fig. 14–2 for the two positions of the detector in two different linear scans. Two columns perpendicular to each other have only a small area in common. This area is called a pixel and the entire cross section can be imagined to consist of these small pixels. Thus different combinations of pixels but always perpendicular to the direction of scan contribute to the counts of the detector at different positions along a scan line at different angles. These are called projections of a plane on a line. From these data then, the radioactivity of each pixel is computed using the method described in the next section. The number of pixels in a cross section is determined by the number of columns, N, in a scan direction and is simply $N \times N$. In the case of Fig. 14–2, this number is $9 \times 9 = 81$. In actual practice in nuclear medicine, the matrix size is either $64 \times 64$ or $128 \times 128$ resulting in 4096 or 16384 pixels respectively.

Pixel width ($\Delta x$), number of pixels N along a scan line, and number of linear scans M at equally spaced angles are three important, interrelated considerations in data acquisition as all three pertain to the amount of time required to complete a scan. Pixel width determines the resolution obtainable in the tomograph which is $2 \times$ the pixel width. Reducing the pixel width, $\Delta x$, in general improves the resolution, while increasing the pixel width

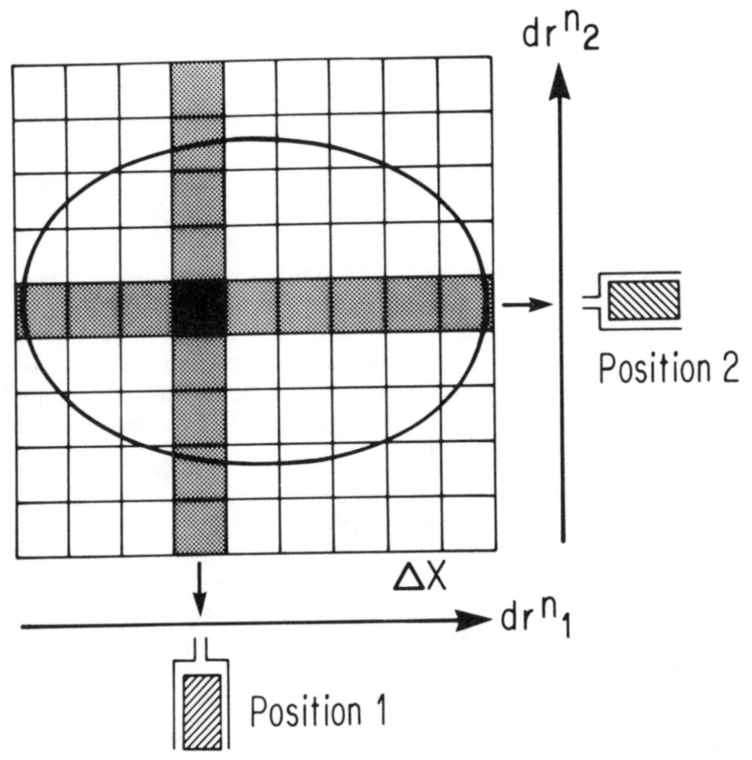


**Fig. 14–2.** Data acquisitions in transverse tomography. $dr^{n1}$, $dr^{n2}$ stand for scan directions 1, 2, . . . .

degrades it. However, to reduce the pixel width $\Delta x$, one has to reduce the size of the hole in the collimator of the detector which reduces the sensitivity of the detector. Thus to achieve the same statistical reliability, one has to increase the time spent at each location. In addition, the reduction in $\Delta x$ increases the number of pixels N along the scan line which necessitates more linear scans M at different angles for optimum sampling. This again requires more time. Thus, we face a similar dilemma in transverse tomography as we face in areal scanning, *i.e.*, improvement in resolution costs dearly in terms of sensitivity and a compromise has to be made between the two. In CT, where, because of the large number of photons available from an x-ray tube, the typical resolution or pixel size is about 1 mm, the number of linear scans performed is 180 or more. In nuclear medicine, where the available number of photons in a typical cross section is relatively small, typical resolution is in the range of 1 cm and the number of linear scans needed is in the vicinity of 64 (sometimes 128).

Besides the width $\Delta x$ of the columns, the other important requirements of collimation are:

1. An individual detector at a given location picks up counts from only one column (perpendicular to the scan direction).

2. There are no variations in counts detected from a uniform cross section at different locations along scans from multiple directions.
3. The same activity in any pixel in a column contributes equally to the counts at a given location, *i.e.*, the detector response should be uniform with depth.
4. All counts come from only the cross section under consideration and not from adjacent cross sections.

These requirements are not easily achieved in nuclear medicine, particularly for SPECT.

A major difference between CT (or PET) and SPECT in data acquisition is that, in CT, data are acquired within a 180° rotation of linear scans whereas, in SPECT, with a few exceptions, a complete rotation (360°) of linear scans around the cross section is utilized. The reason for this difference will be discussed under SPECT.

***Reconstruction of the Cross Section:*** Reconstruction of a cross section from its multiple linear projections around the cross section is a general problem which is encountered in diverse fields. Its solution was provided by Radon as far back as 1917. However, because of the complexity involved in the computations, practical realization of the solution was achieved only in recent years with the advent and easy access to large digital computers. There are a number of mathematical techniques to solve this problem, the most common method is known as filtered back projection which is a modified form (therefore more accurate) of back projection. Back projection can be easily understood with the help of Fig. 14–3.

Let us consider a simple cross section containing two radionuclide sources as depicted in Fig. 14–3a. When data are acquired from this cross section with an appropriate detector and by linear scanning from multiple directions, the responses of the detector are shown as Scan 1, Scan 2, Scan 3, Scan 4, etc. Notice that in this case, the detector receives counts only at one or two discrete locations in each scan and these locations are determined by the intersection of the scan directions and the perpendicular lines from each source to the scan direction respectively.

In the back projection method of reconstruction, as the name implies, data received at one location of a linear scan is back projected in the column right above that location and so on. Since there is no prior knowledge about which or how many pixels in that column contain radioactivity (actually that is what we would like to know), one assumes that counts received at that location came in equal amounts from each pixel in that column. This is done in Fig. 14–3b. Counts received by the detector in Scan 1 at the two locations are distributed equally in the 12 pixels of columns 3 and 7 respectively. When a similar procedure is repeated for other scans (2, 3, 4, etc.), the reconstruction of the cross section is complete. As can be seen two radioactive sources have been reproduced at the correct locations, though not as point sources but as star patterns. As a result, a number of pixels (shown shaded) which contained no radioactivity in the original cross section, are reconstructed inaccurately as containing small amounts of radioactivity. In a more refined approach, known as filtered back projection, the star artifacts are

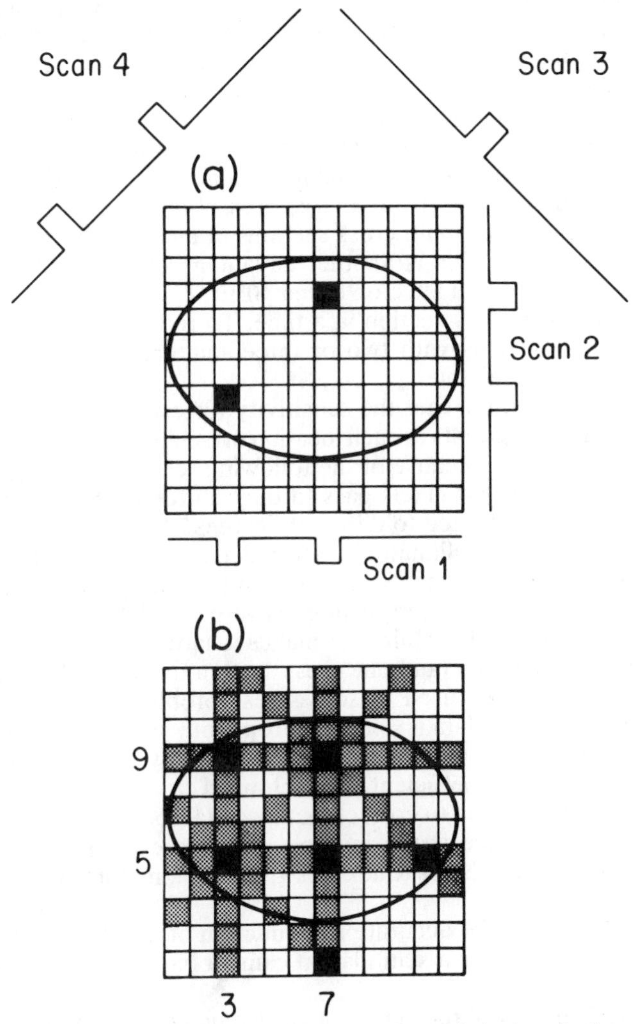


**Fig. 14–3.** Reconstruction of a cross section using simple back projection method.

removed by adding and subtracting fixed fractions of the counts from the column under consideration to several adjacent columns on both sides. When this is done, an accurate and artifact free reconstruction is obtained. In actual practice, data acquired are large and the mathematical operations needed for reconstruction are huge, therefore a large digital computer is essential to store and process the data quickly.

## SINGLE PHOTON EMISSION TOMOGRAPHY (SPECT)

Of the two current approaches for performing SPECT, using either multidetectors or a scintillation camera, the latter is more popular as most of the equipment needed, *e.g.*, a scintillation camera and a computer, is already available in nuclear medicine departments. Only a rotating gantry on which a scintillation camera head can be mounted and a software program to acquire data and reconstruct the cross sections, are additionally needed. In this set up, a scintillation camera head rotates in a circle around a patient, making a number of stops for a given time to acquire data from multiple directions (32 to 64 in routine procedures). Since a scintillation camera acquires data simultaneously from a large area of an organ, as opposed to a thin section, it produces multi cross sections of an organ from just one data acquisition sequence. This is another advantage of a scintillation camera based SPECT over multidetector based SPECT systems which presently reproduce only one cross section at a time. To reduce the acquisition time further, SPECT systems with two or three scintillation-camera heads are now commercially available.

The two requirements, that the sensitivity for each pixel in a column be the same and that the counts at each location originate only from that column, are not met in SPECT because of the following reasons. First, gamma rays originating from different pixels pass through different thickness of tissues and therefore are attenuated to different degrees. Also, even without attenuation, sensitivity of a collimator is depth dependent. Thus, the sensitivity for each pixel is not the same. Secondly, compton scattering of gamma rays originating outside the column of interest into the detector as well as the diverging field of view of a collimator makes it hard to collect data at a given location from one column (and one cross section) only (see Fig. 10–2). Several methods have been used to solve these problems. In one method of attenuation correction, data is collected from both sides of the column (360° rotation instead of 180°) and then, knowing the body contour and assuming a uniform attenuation coefficient for each pixel, a correction matrix is obtained. This matrix is used to correct the raw data from which reconstruction is performed. To reduce scattering and divergence of field view, special collimators with longer holes (thickness L of a collimator) have been found useful.

Other requirements for accurate reproduction of a radioactive cross section which are unique for a scintillation camera based SPECT are:

1. Proper alignment of the center of electronic position (x, y) determining circuit with the center of rotation. In areal imaging, a shift by one or two pixels in x or y direction is hardly noticeable but in SPECT such a misalignment produces recognizable artifacts, *e.g.*, a point source is reconstructed as a ring. Proper alignment can be achieved by placing a point source at the center of rotation, imaging it from two opposite directions and determining the pixel location of the source in the two images. If these are identical, then there is no need for adjustment and the electronic center is properly aligned with the center of rotation. However, if the pixel location of the source is not identical, these are made identical by electronically off setting x or y voltage, as the case may be.
2. Uniformity of response of scintillation camera. This requirement is much more stringent in SPECT than routine imaging as errors prop-

agate rapidly in reconstruction of the image. Whereas in areal imaging uniformity in the range of $\pm 5\%$ is acceptable, in SPECT it should be under $\pm 2\%$. As explained earlier (p. 177) newer scintillation cameras are capable of such a performance.

3. Invariance of uniformity at different angles. Since images are acquired at a large number of angles, the uniformity of scintillation camera should be maintained at each angle. The main reason for changes in the uniformity response at different angles is the change in photomultiplier tube gain which occurs when it is oriented at different angles in earth's magnetic field. In newer cameras where PM tubes are shielded by $\mu$-metal, these effects are reduced considerably.
4. Detector head alignment with the axis of rotation.

Thus clinical success of SPECT depends upon a stringent quality control program, acquisition of sufficient number of counts in each image, use of properly designed collimators, proper correction for attenuation, and, if possible, correction for scatter.

With proper attention to these points, good quality reconstructed SPECT images can be realized routinely in a clinical nuclear department (see Fig. 14–4). However, these images, even though useful clinically, still fall short where accurate quantitative measurements are required.

## POSITRON EMISSION TOMOGRAPHY (PET)

Most of the radionuclides used in nuclear medicine (*e.g.*, $^{99m}Tc$, $^{67}Ga$, $^{201}Tl$) are not isotopes of physiologically important elements such as hydrogen, carbon, nitrogen or oxygen. The only gamma emitting isotopes of these elements are actually positron emitters ($^{11}C$, $^{13}N$ and $^{15}O$) and have short half-lives (in minutes). Fluorine 18 which is also a positron emitter can replace hydrogen in many important biological molecules without changing their function significantly; for example $^{18}F$ deoxyglucose can be used to measure glucose utilization rate by a tissue such as brain. Because of the potential of positron emitting radionuclides for measurement of important physiologic parameters noninvasively, there is a great interest in these radionuclides, even when their short half-life poses a tremendous hardship in the routine availability of radiopharmaceuticals labeled with these radionuclides. Some of these problems in developing short-lived radiopharmaceuticals for use with PET have been recently mitigated with the introduction of turnkey cyclotrons and automated radiopharmaceutical production. In addition, several generator systems specifically for PET use are commercially available, e.g., $^{68}Ge$- $^{68}Ga$ and $^{82}Sr$- $^{82}Rb$.

***Principle:*** Positron emitting radionuclides as such do not emit gamma rays but when an emitted positron has lost its energy by interaction with surrounding medium within a short distance of the site of emission, it annihilates by combining with an electron and two gamma rays with energies of 511 keV each are produced simultaneously. Such a pair of gamma rays produced by annihilation of a positron always travel in opposite (180° apart) directions.

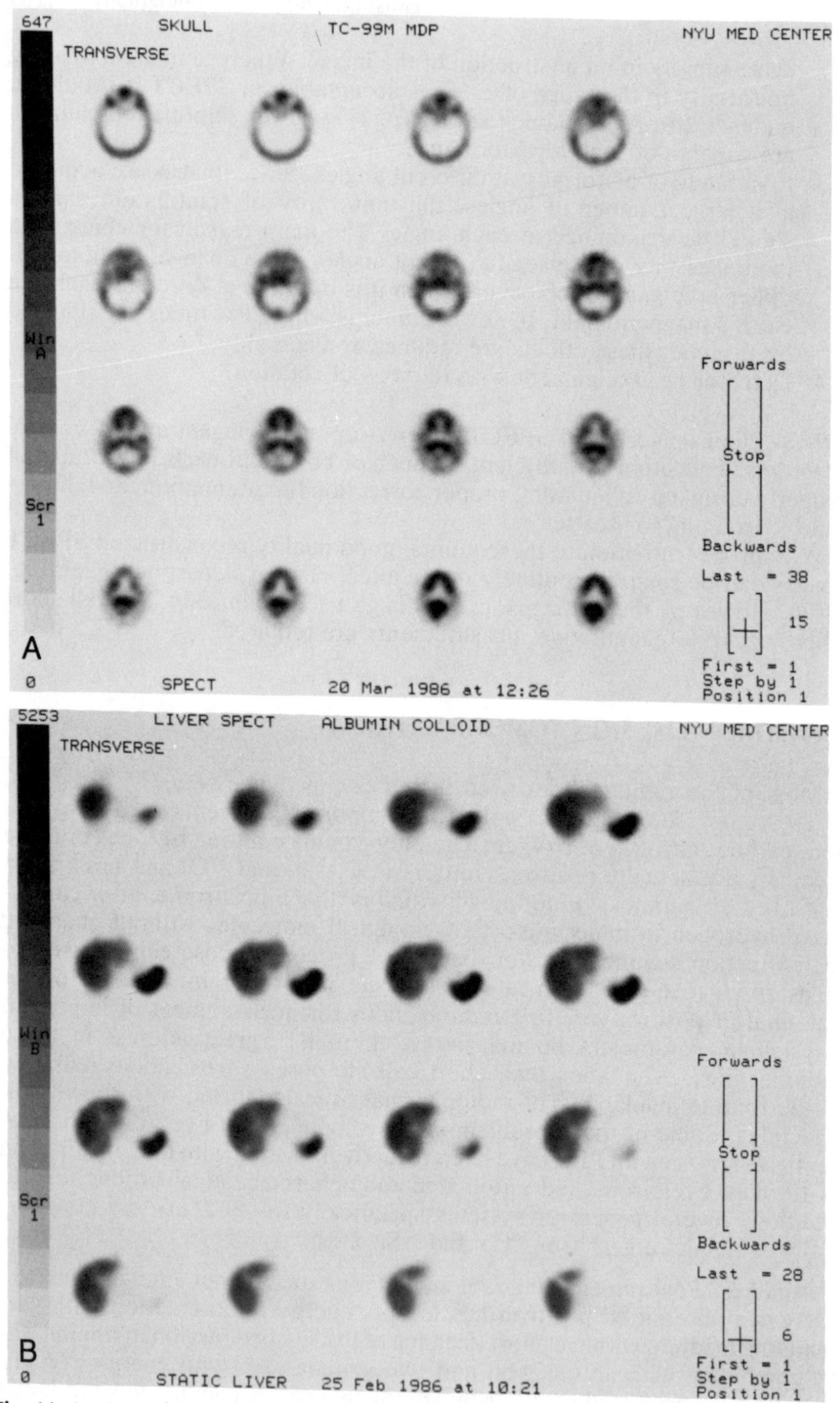


**Fig. 14–4.** Typical SPECT images. (a) 16 transverse cross section through the head showing distribution of $^{99m}$Tc MDP in head and facial bones, and (b) 16 transverse cross sections through the abdomen showing distribution of $^{99m}$Tc albumin aggregates in liver and spleen. (Courtesy of Dr. J Sanger, N.Y.U. Med. Center.)

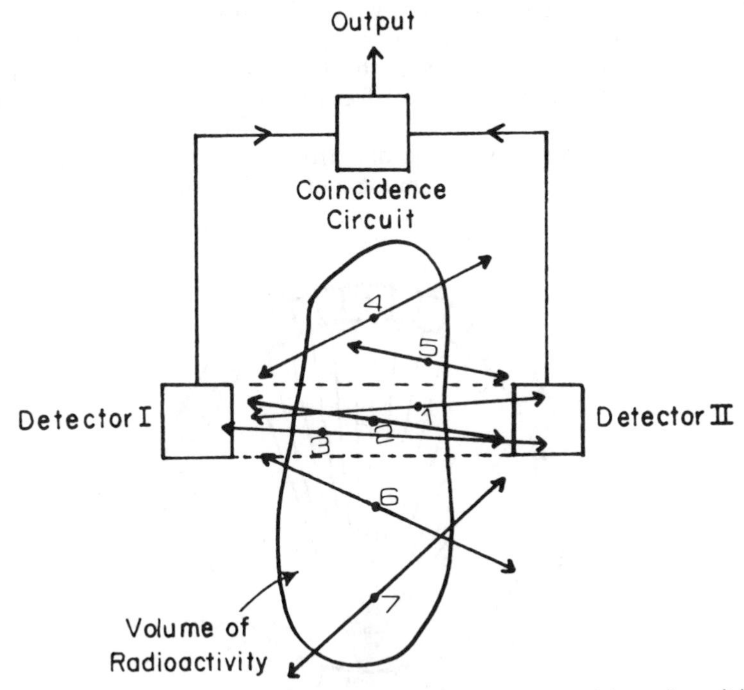


**Fig. 14–5.** Coincidence mode of scanning a distribution of a positron-emitting radionuclide. Only those annihilation photons produced in the volume shown by the dotted lines (e.g., 1, 2 and 3) have a chance of arriving at the two detectors simultaneously. Annihilation photons produced at locations outside this volume (e.g., 4, 5, 6 and 7) will not reach the two detectors simultaneously. Thus, such a pair of detectors measures the radioactivity, without the use of a collimator, of a small volume only. By moving such a pair of detectors over the organ of interest, a scan of the radioactive distribution is produced.

In positron emission tomography or positron scanning, these two gamma rays are used in a coincidence mode where both of the gamma rays are detected by a pair of NaI(Tl) detectors simultaneously as shown in Fig. 14–5. Two small NaI(Tl) detectors face each other on either side of a radioactive organ. Output of each detector is then relayed to a coincidence circuit to determine if the output of each detector originated simultaneously or within a short time of each other ($<10^{-9}$ secs). As can be seen, simultaneous detection of the two annihilation gamma rays defines a nice field of view (a cylindrical column as opposed to a cone for a single detector with a collimator) in a unique manner without the aid of a collimator. One complication that defeats this function arises due to the accidental coincidences that occur from the accidental interactions of two unrelated (not from the same annihilation) rays. Accidental coincidences can be easily controlled as well as corrected. Also note that in this setup, the sensitivity and resolution in the field of view of such a pair of detectors does not vary with the location of radioactive source in its field of view. This, as was stated earlier, is an important requirement for ECT which is difficult to achieve in SPECT but is fulfilled easily in PET. Also the two annihilation gamma rays taken together pass through the same thickness of tissue irrespective of location of the annihilation in the field of view of the detector pair. Therefore, it is much easier and more accurate to correct for attenuation. Again, one more advantage over SPECT.

To perform PET, one can acquire data from a thin cross section with such a pair of detectors by linearly scanning a cross section from multiple direc-

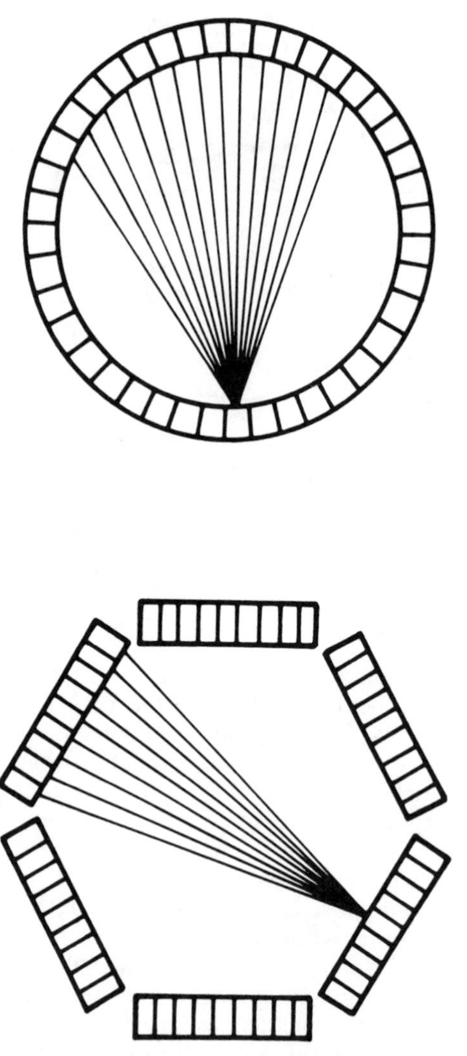

**Fig. 14–6.** Circular or hexagonal arrays of detectors for simultaneous collection of data from various directions in a PET imager. Each detector in the array or ring is in coincidence with a large number of detectors on its opposite side. Here, only one detector is shown in coincidence with several detectors on the opposite side.

tions. However, the sensitivity of such a device will be low. To increase the sensitivity, multiple pairs of detectors either in a ring or a hexagonal array are used in commercially available systems as shown in Fig. 14–6. To increase further the sensitivity of these devices, more than one array or ring is sometimes used to acquire data from multiple cross sections simultaneously. BGO crystals, which have higher intrinsic efficiency for 511 KeV gamma rays, are used instead of NaI(Tl) crystals. The resolution of these devices can approach 5 mm.

Since PET, besides the instrumentation just described, uses short-lived radionuclides, a small on-site cyclotron is needed which makes this technique quite expensive. As a result, only a small number of medical centers are presently exploring the clinical potential of this technique.

## LONGITUDINAL TOMOGRAPHY

In longitudinal tomography, an organ is scanned with the detector head moving in a plane parallel to the plane of interest in the organ. A number of scans are obtained from multiple directions in that plane (plane of detector head motion) by angling the detector head of a rectilinear scanner. These scans are then superimposed in such a way that best resolution is achieved for a given plane in the organ (see Fig. 14–7). Although superimposition can be achieved photographically, use of a computer for this purpose is faster and more reliable. Longitudinal tomography can also be performed with an Anger camera using a collimator with slanted holes and imaging the organ from different directions by rotating the collimator accordingly (Fig. 14–8). A recent addition to this type of tomography is a seven pin-hole collimator which produces seven images of an organ taken from seven different angles. From these images, different planes of the organ are reconstructed. Lon-

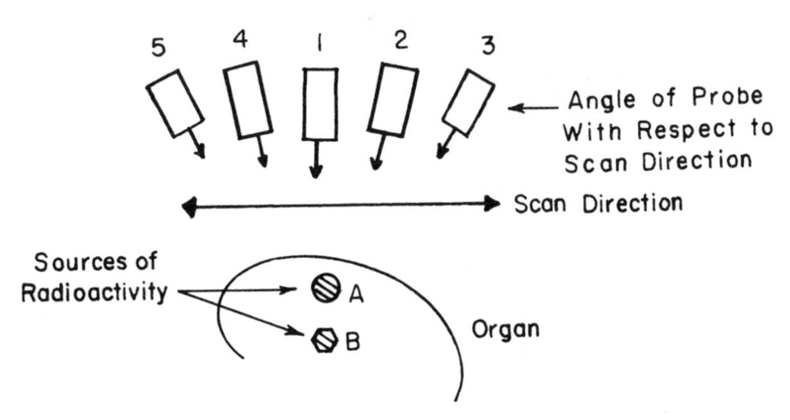


**Fig. 14–7.** Longitudinal tomography using a rectilinear scanner. An organ is scanned several times in a plane parallel to the plane of interest in the organ with the detector head inclined at various angles. The superimposition of these scans in a specified manner produces the tomograph of the plane of interest.

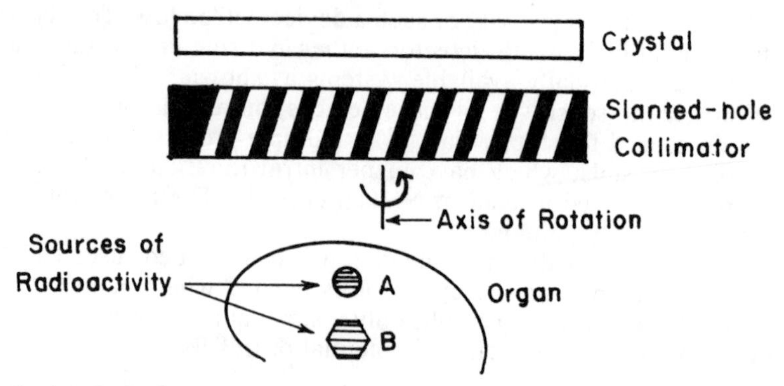


**Fig. 14–8.** Longitudinal tomography using an Anger camera. A slanted parallel-hole collimator is used to image an organ. By rotating this collimator along the axis passing through the center of the collimator and perpendicular to it, multiple images of the organ are obtained from many angles. The reconstruction of the plane of interest is then achieved by superimposition of these images again in a specified manner.

gitudinal tomography has met little clinical success primarily because it does not remove the radioactivity present above or below the plane of interest; it only blurs that information.

## PROBLEMS

1. Is it essential to acquire data from all sides (360° angle) for accurate reconstruction in SPECT? Can it be less than 180°?
2. What is the primary benefit in using a dual or triple head scintillation camera in SPECT? What additional problems, if any, do these entail?
3. Why is quality control important in SPECT?
4. What two factors make quantitation with SPECT less accurate than PET?
5. Why is the resolving time of a coincidence circuit an important parameter in a PET scanner?

# 15

# Biological Effects of Radiation

Interaction of radiation with biological systems results in a variety of biological changes which can be either deleterious or benign. These changes may become evident immediately or may take years, or generations, before being manifested. In general, the probability of occurrence, type, and severity of these changes depends on many factors, some related to radiation and its characteristics, and others to biological characteristics of the system. Since a detailed account of these changes and the factors affecting them fall under the discipline of radiation biology, the discussion here is limited to only those aspects of radiation biology which are essential for an assessment of the possible dangers in the use of radiation as these relate to nuclear medicine.

## MECHANISM OF BIOLOGICAL DAMAGE

Manifestation of bio-injury due to radiation is always preceded by a complex series of physiochemical events shown in Fig. 15–1. The first step in this series of events is the deposition of energy by radiation in the form of ionization and excitation of some of the atoms or molecules of the biological system. This generally lasts about $10^{-12}$ seconds or less. The second step is the transfer of energy either to neighboring molecules (inter-molecular) or, quite often, within the molecule itself (intramolecular) to form various short-lived and chemically active species known as free radicals. This process may last from $10^{-12}$ to $10^{-3}$ seconds. In the next stage the free radicals react either among themselves or, more significantly, with biomolecules (*e.g.*, DNA, RNA) to produce alterations in them. This process may last from milliseconds to several seconds. The final step, *i.e.*, the expression of the biological alteration produced in the last stage, is the biological damage which is tied to the fate of these altered biomolecules. Eventual biological damage may be manifested within a short time or may be delayed up to several generations, depending on the type and function of these altered molecules.

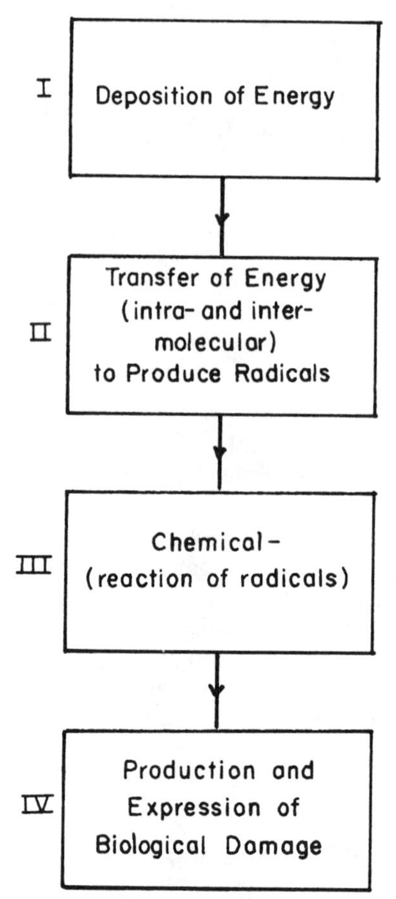


**Fig. 15–1.** Various stages in the development of bio-injury caused by high-energy radiation.

## FACTORS AFFECTING BIOLOGICAL DAMAGE

Biological damage of a system caused by radiation depends on the following factors:

***Radiation Dose:*** Any biological effect of radiation, whether deleterious or benign, strongly depends upon the radiation dose. Generally, more effects, and more serious effects, are produced by high than by low doses. The exact relationship of the dose to the effect produced, however, depends on the nature of the effect. For example, the dose-effect relationship for the induction of cancer differs from that causing genetic mutation.

An important question in this regard is whether there exists a threshold level of radiation dose below which no deleterious effects are produced. Unfortunately, this question is unresolved, particularly in the case of two important effects—the induction of cancer and the production of genetic damage. In the absence of a definite answer to this question, it is generally assumed that dose and effect are linearly related even at low doses, and that there is no threshold radiation dose below which these effects are not produced.

***Dose Rate:*** If the same dose is delivered to two identical biological systems, one with a short duration (high dose rate) and the other over a longer period of time (low dose rate), the biological responses of the two systems will differ. High dose rates are more damaging than low dose rates.

***LET or Type of Radiation:*** Higher LET (see p. 73) radiations ($\alpha$ particles, protons and neutrons as a result of the production of recoil protons) produce greater damage in a biological system than lower LET radiations (electrons, and $\gamma$- and x-rays as a result of compton- and photoelectric interaction). The relative biological effectiveness (RBE) of a radiation for producing a given biological effect is defined as follows:

$$\text{RBE} = \frac{\text{Dose of a standard (x-ray) radiation needed to produce a biological effect}}{\text{Dose of a second radiation needed to produce the same biological effect}}$$

For example, the RBE of 10 MeV neutrons for killing a cell is ~10. In other words, 10 MeV neutrons are ten times more effective at cell killing than x-rays (generally, 250 kVp x-rays are used as a reference).

***Type of Tissue:*** Biological response of a system varies widely depending on the type of tissue (*e.g.*, liver, bone marrow, or nerve tissue) involved. Given the same radiation dose and dose rate, bone marrow is much more sensitive than nerve tissue to certain types of radiation damage.

***Amount of Tissue:*** Injury to a biological system also depends upon the amount of tissue irradiated. For example, a mammal can tolerate a much higher dose to a part of the body compared to irradiation of the total body.

***Biological Variation:*** The response of a biological system, even with all other factors constant, may vary enormously even among closely related individuals. One person may tolerate up to 1,000 rads, whereas a dose of only 200 rads in another person may be lethal.

***Chemical Modifiers:*** The presence of certain chemicals is known to modify the response of a biological system to radiation. Substances which make biological systems more radio-resistant are called radio-protectors. A good example of a radio-protector is the protein, cysteine. Substances which make biological systems more radio-sensitive are called radio-sensitizers. An example of a radio-sensitizer is molecular oxygen. For a radio-protector or radio-sensitizer to have any effect, it must be present in the biological system during irradiation.

## DELETERIOUS EFFECTS IN MAN

Deleterious effects in man may be acute or late. Acute effects are manifested within a short period after irradiation and range from transient nausea

**Table 15–1. Radiation Doses and Acute Effects**

| Stage | Dose Range Rads (Gy) | Symptoms |
|---|---|---|
| I | 0–200 (0–2) | Usually unobservable |
| II | 150–400 (1.5–4) | Transient nausea and vomiting; some evidence of damage to hematopoietic system, recovering in 1 to 2 months |
| III | 350–600 (5–6) | Severe damage to hematopoietic system; bone marrow transplant essential; survival chances, moderate |
| IV | 550–1,000 (5.5–10) | Gastrointestinal damage; severe nausea, vomiting and diarrhea; very small chance of recovery; death follows in 10–24 days |
| V | 1,000 and above (10 and higher) | Confusion, shock, burning sensation; death follows within hours |

and vomiting, to death. Late effects may take as long as several generations for eventual manifestation. These include the induction of cancer and leukemia, birth defects and other abnormalities in the offspring of the irradiated person (genetic damage), cataracts, shortening of the overall life span, and production of temporary or permanent sterility in an individual.

### Acute Effects

These are generally produced when the radiation dose is high and delivered to the total body in a short duration. Five clinically distinct stages can be identified as the radiation dose is progressively increased: (1) no effect; (2) mild damage to bone marrow; (3) severe damage to bone marrow and mild damage to the gastrointestinal tract; (4) severe damage to gastrointestinal tract; and (5) damage to nerve tissue. Table 15–1 lists the radiation doses and the typical symptoms produced for these five stages.

### Late Effects

Late effects may be expected in cases where acute reactions are minimal (*i.e.*, when the radiation dose to the total body is low or only part of the body is involved in the irradiation). In diagnostic uses of radiation, whether in radiology or nuclear medicine, the range of radiation doses generally delivered to a patient falls in this category. Therefore, the dose-effect relationship of late effects is of special interest.

Regrettably, precise information about the risks involving low radiation doses is difficult to obtain for the following reasons:

1. The probability of occurrence of any late effect after low doses is small. Therefore, to perform statistically valid studies, large populations have to be considered. In practice, this is difficult to implement.

2. The occurrence of a latent period in the expression of late effects requires long follow-ups (10 years or more).

3. Late effects of radiation also occur naturally. Since accurate information is lacking on the natural frequency of these effects, it is difficult to estimate the influence of low radiation doses on this increase. In addition, the frequency of natural occurrence is influenced by many complex factors

**Table 15–2. Excess Cancer Mortality—Life Time Risk, Single Exposure**

| *Type* | *Male (per rad per $10^6$ Persons)* | *Female (per rad per $10^6$ Persons)* |
|---|---|---|
| Leukemia | 660 | 730 |
| Nonleukemia | 110 | 80 |
| Total | 770 | 810 |

such as age, sex, genetic history, geography and various environmental and socioeconomic factors.

In spite of these obstacles, a fair estimation has been made of the risks to a person or a population at low radiation doses. Most of the information in this regard is derived from two sources:

1. Extrapolation from experiments conducted on laboratory animals (*e.g.*, mice).

2. Retrospective studies of the victims of the atomic bomb explosions in Nagasaki and Hiroshima, the inhabitants of Bikini and other Pacific islands who were exposed to radiation as a result of fall-out, persons irradiated for medical reasons, and occupational workers such as uranium mine workers and radiation workers (*e.g.*, radiologists) who are exposed to radiation as a result of their occupations.

Of all the late effects, the induction of cancer and genetic damage are the most important. Table 15–2 lists the risk of cancer as deduced from the currently available information.* Information on genetic damage has been deduced mostly from results obtained in experiments with Mediterranean fruit flies (*Drosophila*) and laboratory mice. The present estimate of the doubling dose for genetic effects (the dose needed to double the natural incidence of a genetic or somatic anomaly) is about 100 rads (1 Gy). It is now generally accepted that carcinogenic effects of radiation are more significant than the genetic effect.

## RADIATION EFFECTS IN A FETUS

The embryo and fetus are more susceptible to radiation than the adult. Radiation during fetal life may not only cause fetal death and induce cancer, but also may produce various gross and behavioral abnormalities and malformations. The most sensitive stage for the production of these abnormalities is between the second and tenth weeks of pregnancy (first trimester). Radiation studies of rodents during this interval suggest that doses as low as 5 rads may produce gross congenital malformations. Below 5 rads the main risk to a fetus is the increased probability of cancer, which at present

* Health effects of exposure to low levels of ionizing radiation, BEIR Report, Division of Medical Sciences. National Academy of Sciences, National Research Council, National Academy Press, Washington, D.C., 1990.

is estimated to be about 500 deaths before 10 years of age among one million children exposed shortly before birth to 1 rad. These are the excess deaths, not the total number of cancers induced.

## PROBLEMS

1. List the various stages in the development of radiation injury.
2. For the same radiation dose, what are the other factors that can modify the biological effects or their intensity?
3. Generally, below what dose are acute effects not observable?
4. Name the most common late biological effect and give its probability of occurrence after 1 rad of total body irradiation.
5. What is the most sensitive period during pregnancy for the development of a fetal abnormality?

# 16

# Safe Handling of Radionuclides

Any use of radionuclides is fraught with the danger of inadvertently exposing an individual to radiation and, therefore, to its attendant hazards. This is especially true in the nuclear-medicine laboratory where large amounts of unsealed radioactive sources are routinely handled. Each time a generator is milked, a radiopharmaceutical dose is drawn or injected, or a scan is performed on a patient, there is the possibility of exposure and contamination to the user (technician, physicist or physician) as well as to the environment. To ensure proper and safe use of radionuclides, a number of Federal, State and local agencies regulate production, transportation, possession, use and disposal of radionuclides. Some of the Federal regulations are published in the code of Federal Regulations under Title 10. The reader is urged to read this document, particularly parts 20 and 35 which are of direct relevance to nuclear medicine. Table 16–1 lists the exposure limits or maximum permissible dose (MPD) to various categories of exposed persons. These are legal limits which are not to be exceeded at any time. These limits do not imply that radiation below this level is harmless or safe. The guiding principle in the radiation protection or health physics is that radiation doses should be "as low as reasonably achievable" (ALARA). This philosophy is now mandated by law and therefore puts the onus on the user to reduce radiation levels in work places to as low a level as is economically and technologically feasible, even if these levels are well below the legal limits. Keeping this in mind, this chapter focuses on some practical ways in which an individual user can minimize exposure to oneself and fellow workers, and reduce contamination of the environment.

## REDUCING EXPOSURE FROM EXTERNAL SOURCES

Hazard from external radiation sources is measured in terms of exposure. Exposure is a measure of the ability of radiation to produce ionization in air. The unit of exposure is the roentgen (R), that level of radiation which produces ionization in air in the amount of $2 \cdot 58 \times 10^{-4}$ coulomb/kg of air. A milliroentgen (mR) is one thousandth of a roentgen. The SI units of exposure are simply coulomb/kg. Therefore 1 coulomb/kg = 3876 R. If the

**Table 16–1. Yearly Maximum Permissible Dose (NRC 1990)**

| | |
|---|---|
| Radiation Workers | |
| Whole body | 5 rem* or 50 mSv* |
| Hands, forearms, etc. | 75 rem or 750 mSv |
| Skin of the whole body | 30 rem or 300 mSv |
| Pregnant Radiation Worker (fetus) | 0.5 rem or 5 mSv (Single exposure) |
| Non-Radiation Worker or Public | 0.1 rem or 1 mSv** |

* rem = radiation dose in rads × quality factor. Quality factor for radiations encountered in nuclear medicine equals one. Sievert (Sv) replaces rem in SI units and 1 Sv = 100 rem or 1 rem = 10 mSv.
** To be effective from January 1993. Old limits are 0.5 rem or 5 mSv.

exposure is known at a given point, one can calculate the absorbed dose to a person standing at that point by multiplying the exposure by a term known as "f" factor. For muscle and soft tissue, f factor is close to unity and, therefore, for our purposes in nuclear medicine, we may assume that exposure is more or less equivalent to the absorbed dose.

The principle of reducing exposure to radiation from sources outside the body (x- and γ-ray emitting radionuclides only) can be summed up in three equally important words: *Time, Distance* and *Shielding.*

***Time:*** Since the total radiation dose, whether to the total body or a part of the body, is directly proportional to the time of exposure, it is extremely important to spend as little time as possible near x- or γ-ray-emitting radionuclide sources. This requires forethought and precaution by the user. For example, when eluting a $^{99}$Mo generator, one should not stand near it while the elution is in progress. When injecting a dose to a patient, one should locate the vein first, and then take the syringe containing the radionuclide dose out of the leaded syringe carrier. However, rushing through a procedure is definitely not suggested here—if one has to repeat a procedure because of haste, total exposure time will be doubled.

***Distance:*** Radiation dose to the body from a small external source varies inversely with the square of the distance of the source from the body (*i.e.*, if we increase the distance from a source of radiation by a factor of two, the radiation dose will drop by a factor of 4). Therefore, one should use a cart with a long handle when carrying very hot (>50 mCi) radionuclide sources emitting high-energy γ-rays (>300 keV) from one place to another—even though these sources are shielded. When radionuclide sources which are moderate in activity (~10 mCi) are hand-carried, the source should be held away from the body.

In drawing a dose for injection, use a syringe large enough so that it is no more than half-full when the desired volume is added, and handle it from the unfilled area. If it is necessary to work with hot sources for long periods of time, then one should seriously consider the use of various remote-control tools available for picking up a radioactive vial or pipetting a radioactive solution.

***Shielding:*** γ- and x-rays can be effectively shielded using thick containers, bricks or partitions made of lead. Mirrors should be used for viewing behind lead partitions. The amount of actual shielding needed to reduce exposure to a minimal level depends on the amount of radioactivity to be shielded and the energy of the $\gamma$-rays.

The three variables, time, distance and shielding can be combined in a single formula which gives the exposure from a small radioactive source:

$$E = \frac{n\Gamma}{d^2} \cdot e^{-\mu(\text{linear}) \cdot x} \cdot t \qquad (1)$$

Here E is the exposure (R), n is the number of millicuries in the source, d is the distance (cm) of the point (at which exposure is desired) from the source, t is the time of exposure (hr), $\mu$(linear) is the linear attenuation coefficient of the shielding material ($cm^{-1}$), and x is the thickness of the shielding material (cm). Gamma, $\Gamma$,* is the exposure rate constant of the radionuclide and its units are $R \cdot cm^2/mCi \cdot h$. For $^{99m}Tc$, its value is 0.60. For other radionuclides of interest to nuclear medicine, its values are given in Appendix A.

*Examples*

(1) Calculate the exposure in one minute to the tips of the fingers from a syringe held by the tips of the fingers and containing 15 mCi of $^{99m}Tc$ radioactivity (assume the distance of the radioactivity from the tips of the fingers as 3 cm).

In this case,

$$n = 15 \text{ mCi}, \Gamma = 0.60 \text{ R} \cdot \text{cm}^2/\text{mCi} \cdot \text{h}$$

$$t = 1 \text{ min} = \frac{1}{60} \text{ h}$$

$$x = 0 \text{ (no shielding, neglect absorption in the source)}$$

$$d = 3 \text{ cm}$$

Substituting these values in equation (1),

$$E = \frac{15 \times 0.6}{3^2} \cdot e^{-\mu(\text{linear}) \cdot 0} \cdot \frac{1}{60} \text{ R}$$

$$= 0.017 \text{ R} = 17 \text{ mR}$$

(2) Calculate the same exposure as in example 1 except that this time the syringe is shielded by 1 mm thick lead [$\mu$(linear) = 25 $cm^{-1}$]. In this case, x = 1 mm = 0.1 cm and $\mu$(linear) = 25 $cm^{-1}$.

All other factors are the same.

* For radiation protection work, $\gamma$- or x-rays with energies less than 20 keV are not included in the calculation of $\Gamma$.

Therefore, using equation 1, we get

$$E = \frac{15 \times 0.6}{3^2} \cdot e^{-25 \times 0.1} \cdot \frac{1}{60} R$$

$$= 0.017\, e^{-2.5}$$

$$= 0.017 \times 0.08 = 0.00136 R$$

$$= 1.36 \text{ mR}$$

(3) A patient has been injected with 15 mCi of $^{99m}$Tc radioactivity. Calculate the exposure to a technician who, on average, takes 30 minutes to perform the scan and stays at about 1 meter distance away from the patient during this time. (Assume the radioactivity is localized in a small volume in the patient and that there is no attenuation in the patient).
In this case,

$$n = 15 \text{ mCi}, \Gamma = 0.60 \text{ Rcm}^2/\text{mCi}\cdot\text{h}$$

$$d = 1 \text{ meter} = 100 \text{ cm}$$

$$t = 30 \text{ min} = 0.5 \text{ h}$$

$$x = 0 \text{ and } \mu(\text{linear}) = 0$$

Using equation (1)

$$E = \frac{15 \times 0.6}{100^2} \cdot e^{-0} \cdot 0.5 R$$

$$= \frac{15 \times 0.6 \times 0.5}{10{,}000} = 0.00045 R$$

$$= 0.45 \text{ mR}$$

If the attenuation in the patient is also taken into account, the exposure will be further reduced, probably by a factor of 4.

These illustrative examples show the extent of the possible exposure in nuclear medicine. The other and very important source of exposure is the radionuclide generator.

## AVOIDING INTERNAL CONTAMINATION

Internal contamination by a radionuclide is possible by three routes—penetration through skin, ingestion and inhalation. To avoid ingestion or penetration through the skin, the following steps should be taken:

1. Wear coveralls or a laboratory coat and disposable hand gloves each time you handle a radioactive material. Remember that the hand gloves, once you handle a radioactive material, become contaminated. Since they

are the easiest source of spread of contamination in the laboratory, discard them immediately after handling radioactive material. When handling highly radioactive material, it is wise to wear two pairs of gloves.

2. Do not eat, drink, or smoke in the radionuclide laboratory, or pipette radioactive solutions by mouth. Before eating, when you have been handling radioactive compounds, wash your hands thoroughly.

3. Keep the area neat in which radionuclides are handled. Use a tray with absorbent liners on the bench to limit the spread of radioactive material in case of an accident while working with unsealed radioactive sources.

4. Because the bedsheets, pillows and stretchers used when scanning patients may be contaminated as a result of a patient's saliva, blood, or urine, beware of this route of personal contamination.

Contamination by inhalation does not pose a great problem in nuclear medicine except in a few cases where radioactive gases are used or large amounts of radioiodine are handled. To avoid the contamination of the room by radioactive gases, such items should be stored under a hood and, during use, proper care should be taken to prevent these gases from escaping into the room. In the case of large amounts of radioiodine, all work should be performed under a hood if possible. When opening a vial containing high amounts of radioiodine labeled compounds, the vial should be extended away from you. In the administration of therapeutic doses of $^{131}I$, allow the patient to open the vial and drink the dose by himself.

## THE RADIOACTIVE PATIENT

A radioactive patient is an important source of radiation in nuclear medicine. Through penetrating radiations (x or $\gamma$ rays), he or she can irradiate persons at a distance. Through physical contact or excretion and exhalation into the environment, he or she becomes a source of internal contamination. Presently, a patient has to be isolated in a private room if he or she has received more than 30 mCi (111 MBq) of radioactivity of a radionuclide. With radioactivity under this amount, there are no such restrictions, but because of ALARA, proper health physics principles, some of which have been described earlier, should be followed so as to keep the radiation in the department to a minimum level.

A nursing mother is a special case of a radioactive patient that needs to be discussed separately. Since many radionuclides are excreted in milk, an infant dependent on his or her mother's milk may be exposed to undue risk. In such cases, all the alternatives should be carefully weighed. For tests performed with technetium-labeled compounds, a simple rule of two days' cessation of breast feeding works well. For all of these compounds, the mother's radioactivity will decay by a factor of 1000 or more during this interval, and therefore little if any will be ingested by the infant. However, for tests using radionuclides such as $^{67}Ga$, $^{111}In$, $^{131}I$, and $^{201}Tl$, the situation is quite complex. Careful assessment and monitoring, coupled with several weeks of cessation, may be required to make sure that the infant is not exposed to undue risk.

## PROBLEMS

1. Which of the radionuclides with the following exposure rate constants ($\Gamma$) poses the most danger as an external source of radiation? (a) 2.0 R/mCi/hr, (b) 100 mR/mCi/hr, (c) 2.0 R/mCi/d, and (d) 1.0 R/mCi/min.
2. A patient who was treated with 100 mCi of radio-iodine measured 10 mR/h at 100 cm at the time of administration. By the next day, this measurement dropped to 2 mR/h. How much radioactivity does the patient still have?
3. A radioactive syringe produces an exposure rate of 1 R/hr at a distance of 20 cm. Calculate the exposures of two persons who were working for 3 hours at 50 cm and 100 cm, respectively, from the syringe.
4. Which is the larger source of internal contamination, ingestion or inhalation?

# Appendix A

# Physical Characteristics of Some Radionuclides of Interest in Nuclear Medicine

**Table A–1. Radiations Emitted in the Decay of $^{123}I$ ($\Gamma$ = 1·53 R·cm²/mCi·hr); $T_{\frac{1}{2}}$ = 13 hr**

| *Number* | *Radiation (i)* | *Frequency of Emission ($n_i$)* | *Mean Energy (MeV) ($E_i$)* |
|---|---|---|---|
| 1 | $\gamma 1$ | 0.84 | 0.159 |
| 2 | K Conversion Electron | 0.13 | 0.127 |
| 3 | L Conversion Electron | 0.02 | 0.154 |
| 4 | $\gamma 2$ | 0.01 | 0.529 |
| 5 | X-Ray-K ($\alpha$) | 0.71 | 0.027 |
| 6 | X-Ray-K ($\beta$) | 0.15 | 0.031 |
| 7 | X-Ray-L | 0.13 | 0.003 |
| 8 | LMM Auger Electron | 0.92 | 0.003 |
| 9 | MXY Auger Electron | 2.19 | 0.001 |

**Table A–2. Radiations Emitted in the Decay of $^{131}I$ ($\Gamma$ = 2·2 R·cm²/mCi·hr); $T_{\frac{1}{2}}$ = 8.1 days**

| *Number* | *Radiation (i)* | *Frequency of Emission ($n_i$)* | *Mean Energy (MeV) ($E_i$)* |
|---|---|---|---|
| 1 | $\beta 1$ | 0.02 | 0.069 |
| 2 | $\beta 2$ | 0.07 | 0.096 |
| 3 | $\beta 3$ | 0.90 | 0.192 |
| 4 | $\gamma 1$ | 0.03 | 0.080 |
| 5 | K Conversion Electron | 0.03 | 0.046 |
| 6 | $\gamma 2$ | 0.06 | 0.284 |
| 7 | $\gamma 3$ | 0.82 | 0.364 |
| 8 | K Conversion Electron | 0.02 | 0.330 |
| 9 | $\gamma 4$ | 0.07 | 0.637 |
| 10 | $\gamma 5$ | 0.02 | 0.723 |

Data derived from Journal of Nuclear Medicine (Suppl. 10, 1975).

**Table A–3. Radiations Emitted in the Decay of $^{201}$Tl ($\Gamma$ = 0.47 R·cm²/mCi·hr); $T_{\frac{1}{2}}$ = 73 hr**

| Number | Radiation (i) | Frequency of Emission ($n_i$) | Mean Energy (MeV) ($E_i$) |
|---|---|---|---|
| 1 | $\gamma$1 | 0.01 | 0.032 |
| 2 | L Conversion Electron | 0.21 | 0.018 |
| 3 | M Conversion Electron | 0.07 | 0.029 |
| 4 | $\gamma$2 | 0.04 | 0.135 |
| 5 | K Conversion Electron | 0.10 | 0.052 |
| 6 | L Conversion Electron | 0.02 | 0.121 |
| 7 | $\gamma$3 | 0.12 | 0.167 |
| 8 | K Conversion Electron | 0.18 | 0.084 |
| 9 | L Conversion Electron | 0.03 | 0.154 |
| 10 | X-Ray-K ($\alpha$) | 0.78 | 0.070 |
| 11 | X-Ray-K ($\beta$) | 0.22 | 0.081 |
| 12 | X-Ray-L | 0.46 | 0.010 |
| 13 | KLL Auger Electron | 0.03 | 0.055 |
| 14 | KLX Auger Electron | 0.02 | 0.066 |
| 15 | LMM Auger Electron | 0.81 | 0.008 |
| 16 | MXY Auger Electron | 2.44 | 0.003 |

**Table A–4. Radiations Emitted in the Decay of $^{133}$Xe ($\Gamma$ = 0·15 R·cm²/mCi·hr); $T_{\frac{1}{2}}$ = 5.3 days**

| Number | Radiation (i) | Frequency of Emission ($n_i$) | Mean Energy (MeV) ($E_i$) |
|---|---|---|---|
| 1 | $\beta$1 | 0.02 | 0.075 |
| 2 | $\beta$2 | 0.98 | 0.101 |
| 3 | $\gamma$1 | 0.01 | 0.080 |
| 4 | K Conversion Electron | 0.01 | 0.044 |
| 5 | $\gamma$2 | 0.36 | 0.081 |
| 6 | K Conversion Electron | 0.53 | 0.045 |
| 7 | L Conversion Electron | 0.08 | 0.076 |
| 8 | M Conversion Electron | 0.03 | 0.080 |
| 9 | X-Ray-K ($\alpha$) | 0.39 | 0.030 |
| 10 | X-Ray-K ($\beta$) | 0.09 | 0.035 |
| 11 | X-Ray-L | 0.08 | 0.004 |
| 12 | Auger Electrons | 1.67 | 0.003 |

**Table A–5. Radiations Emitted in the Decay of $^{111}$In ($\Gamma$ = 1·9 R·cm²/mCi·hr); $T_{\frac{1}{2}}$ = 67.4 hr**

| Number | Radiation (i) | Frequency of Emission ($n_i$) | Mean Energy (MeV) ($E_i$) |
|---|---|---|---|
| 1 | Gamma 1 | 0.90 | 0.172 |
| 2 | K Conversion Electron | 0.09 | 0.145 |
| 3 | L Conversion Electron | 0.01 | 0.168 |
| 4 | Gamma 2 | 0.94 | 0.247 |
| 5 | K Conversion Electron | 0.05 | 0.220 |
| 6 | L Conversion Electron | 0.007 | 0.243 |
| 7 | K ($\alpha$) X-Ray | 0.70 | 0.023 |
| 8 | K ($\beta$) X-Ray | 0.14 | 0.026 |
| 9 | L-X-Ray | 0.11 | 0.003 |
| 10 | KLL Auger Electron | 0.11 | 0.019 |
| 11 | KLX Auger Electron | 0.04 | 0.022 |
| 12 | LMM Auger Electron | 0.99 | 0.002 |

**Table A–6. Radiations Emitted in the Decay of $^{67}$Ga ($\Gamma$ = 0·80 R·cm²/mCi·hr); $T_{\frac{1}{2}}$ = 78.1 hr**

| Number | Radiation (i) | Frequency of Emission ($n_i$) | Mean Energy (MeV) ($E_i$) |
|---|---|---|---|
| 1 | Gamma 1 | 0.033 | 0.091 |
| 2 | Gamma 2 | 0.38 | 0.093 |
| 3 | K Conversion Electron | 0.28 | 0.084 |
| 4 | L Conversion Electron | 0.038 | 0.092 |
| 5 | M Conversion Electron | 0.013 | 0.093 |
| 6 | Gamma 3 | 0.24 | 0.185 |
| 7 | Gamma 4 | 0.025 | 0.209 |
| 8 | Gamma 5 | 0.16 | 0.300 |
| 9 | Gamma 6 | 0.04 | 0.394 |
| 10 | K-X-Ray | 0.46 | 0.009 |
| 11 | Auger Electron | 0.66 | 0.008 |

# Appendix B

# CGS and SI Units

| Quantity | CGS Units | MKS or SI Units | Conversion Factors* (CGS → MKS) |
|---|---|---|---|
| Length | Centimeter (cm) | Meter (m) | 0.01 |
| Mass | Gram (gm) | Kilogram (kg) | 0.001 |
| Time | Second (s) | Second (s) | 1 |
| Energy | erg | Joule (J) | $10^{-7}$ |
| Radioactivity | Curie (Ci) | Becquerel (Bq) | $3.7 \times 10^{10}$ |
| Radiation Absorbed Dose | rad | Gray (Gy) | 0.01 |
| Exposure | roentgen (R) | Coulomb/kilogram (C/kg) | $2.58 \times 10^{-4}$ |
| Dose Equivalent | rem | Sievert (Sv) | 0.01 |

* To obtain results in the MKS system, multiply the CGS values by the conversion factor. To obtain results in the CGS system, divide the MKS values by the conversion factor.

# Appendix C

# Exponential Table

| $x$ | $e^{-x}$ | $x$ | $e^{-x}$ | $x$ | $e^{-x}$ |
|---|---|---|---|---|---|
| 0.00 | 1.00 | 0.22 | 0.80 | 0.60 | 0.55 |
| 0.01 | 0.99 | 0.24 | 0.79 | 0.65 | 0.52 |
| 0.02 | 0.98 | 0.26 | 0.77 | 0.693 | 0.50* |
| 0.03 | 0.97 | 0.28 | 0.76 | 0.75 | 0.47 |
| 0.04 | 0.96 | 0.30 | 0.74 | 0.80 | 0.45 |
| 0.05 | 0.95 | 0.32 | 0.72 | 0.85 | 0.42 |
| 0.06 | 0.94 | 0.34 | 0.71 | 0.90 | 0.41 |
| 0.07 | 0.93 | 0.36 | 0.70 | 1.00 | 0.37 |
| 0.08 | 0.92 | 0.38 | 0.68 | 1.50 | 0.22 |
| 0.09 | 0.91 | 0.40 | 0.67 | 2.00 | 0.13 |
| 0.10 | 0.90 | 0.42 | 0.66 | 2.50 | 0.08 |
| 0.12 | 0.89 | 0.44 | 0.64 | 3.00 | 0.05 |
| 0.14 | 0.87 | 0.46 | 0.63 | 3.50 | 0.03 |
| 0.16 | 0.85 | 0.48 | 0.62 | 4.00 | 0.02 |
| 0.18 | 0.84 | 0.50 | 0.61 | 4.50 | 0.01 |
| 0.20 | 0.82 | 0.55 | 0.58 | 5.00 | 0.007 |

* See Chapter 3, p. 32

# Suggestions for Further Reading


1. Johns, H.E., and Cunningham, J.R.: *The Physics of Radiology*, Fourth Ed., Springfield, Charles C Thomas, 1983.
2. Sorenson, J.A., and Phelps, M.E.: *Physics in Nuclear Medicine*, Second Ed., New York, Grune and Stratton, 1987.
3. Saha, G.B.: *Fundamentals of Nuclear Pharmacy*, Second Ed., Springer Verlag, New York, 1983.
4. Loevinger, R., et al.: *MIRD Primer for Absorbed Dose Calculations*. New York, Society of Nuclear Medicine, 1988.
5. Craft, B.Y.: *Single-Photon Emission Computed Tomography*. Chicago, Year Book Medical Publishers, Inc., 1986.
6. Beir, V.: Health Effects of Exposure to Low Levels of Ionizing Radiation. Washington, D.C., National Research Council, National Academy Press, 1990.
7. Shapiro, J.: *Radiation Protection—A Guide for Scientists and Physicians*. Second Ed. Cambridge, Mass., Harvard University Press, 1988.

# Index

Page numbers in italics indicate illustrations; numbers followed by "t" indicate tables.